Applied Chemistry: A Textbook

Applied Chemistry: A Textbook

Bob Garner

Larsen & Keller
www.larsen-keller.com

Applied Chemistry: A Textbook
Bob Garner
ISBN: 979-8-88836-041-5 (Hardback)

Larsen & Keller
Published by Larsen and Keller Education,
5 Penn Plaza,
19th Floor,
New York, NY 10001, USA

Cataloging-in-Publication Data

Applied chemistry : a textbook / Bob Garner.
p. cm.
Includes bibliographical references and index.
ISBN 979-8-88836-041-5
1. Chemistry, Technical. 2. Chemistry. I. Garner, Bob.
TP145 .A67 2023
660--dc23

For more information regarding Larsen and Keller Education and its products, please visit the publisher's website www.larsen-keller.com

Table of Contents

Preface

Over the recent decade, advancements and applications have progressed exponentially. This has led to the increased interest in this field and projects are being conducted to enhance knowledge. The main objective of this book is to present some of the critical challenges and provide insights into possible solutions. This book will answer the varied questions that arise in the field and also provide an increased scope for furthering studies.

The field of applied chemistry deals with the application of theories and principles of chemistry to practical problems. It also seeks to understand the basic chemical properties of materials and developing new materials with well-controlled functions. Some of the major specializations in the field of applied chemistry include analytical chemistry, materials design, environmental technology and bioscience. Analytical chemistry makes use of methods and instruments for the purpose of identifying, separating and quantifying matter. It comprises modern instrumental methods and classical wet chemical methods. Environmental technology involves the application of scientific knowledge and engineering to solve environmental problems and address sustainability. This book contains comprehensive coverage of different concepts related to the field of applied chemistry. It is an essential guide for both academicians and those who wish to pursue this discipline further. Coherent flow of topics, student-friendly language and extensive use of examples make this book an invaluable source of knowledge.

I hope that this book, with its visionary approach, will be a valuable addition and will promote interest among readers. Each of the authors has provided their extraordinary competence in their specific fields by providing different perspectives as they come from diverse nations and regions. I thank them for their contributions.

Bob Garner

1

High Polymers and Plastics

1.1 Polymerization, Introduction and Mechanism of Polymerization

Polymers

Polymers are macro molecules that are formed by repeated linking of small molecules known as monomers.

Example: Nylon, PVC, Polyethylene.

Monomers

Monomers are micro molecule which combines with each other to form a polymer.

Example: $CH_2 = CH_2$.

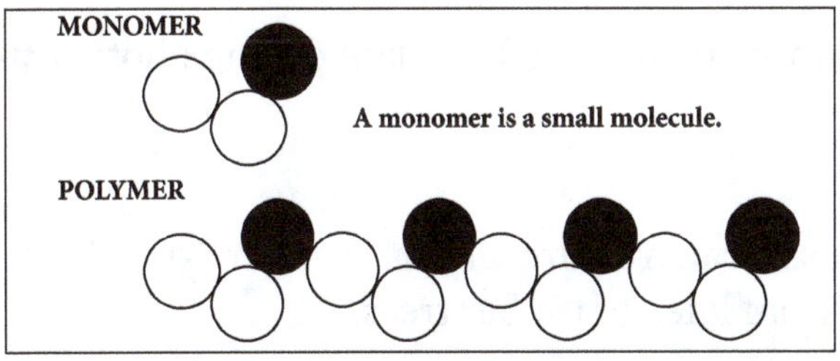

Structure of monomer and polymer.

Classification of Polymers

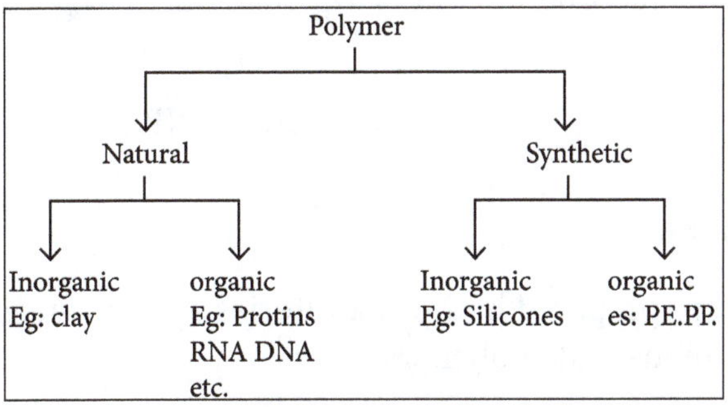

Types of Polymer.

Depending upon the source from which they are obtained, polymers are classified into two types. They are:

- Natural polymers.
- Synthetic polymers.

Types

Based on the chemical composition, polymers are classified as follows:

- Organic polymer.
- Element organic.
- Inorganic polymers.

Natural Polymers

The polymers that are obtained from plants and animals are called natural polymers.

Example:

- Starch: It is a polymer of glucose. It is a chief food reserve of plants.
- Cellulose: It is a polymer of glucose. It is a chief structural material of the plants.

Both starch and cellulose are produced by plant during photosynthesis.

Natural Rubber

It is a polymer of unsaturated hydrocarbon, 2-methyl-1, 3 butadiene, known as isoprene. It is obtained from latex of rubber trees.

$$nCH_2{=}\underset{\displaystyle CH_3}{\underset{|}{C}}-CH=CH_2 \xrightarrow{\text{polymerisation}} (-CH_2-\underset{\displaystyle CH_3}{\underset{|}{C}}=CH-CH_2-)_n$$

Isoprene Polysioprene (NR)

Synthetic Polymers

The polymers which are prepared in the laboratories are known as synthetic polymers. These are also called man-made polymers.

Example: PVC, polyethylene, Teflon, Bakelite, etc.

1. Organic Polymer

These are polymers containing hydrogen, oxygen, nitrogen, sulphur and halogen atoms apart from carbon atoms.

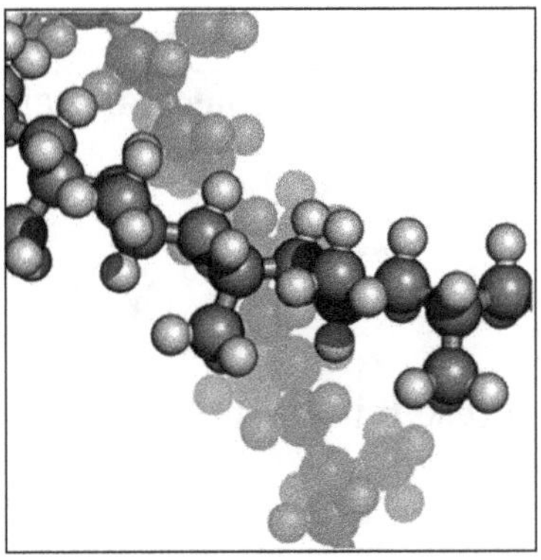

Synthetic organic polymers.

Example: PVC (Polyvinyl chloride), Epoxy polymer, polyurethane.

2. Element Organic

These are polymers composed of carbon atoms and heteroatoms like N, S and O. The main chain consists of carbon atoms and side groups contains hetero atoms linked directly to the 'c' atoms in the chain.

Example: Polysiloxane.

3. Inorganic Polymers

These are polymers containing no carbon atoms. The chains of these polymers are composed of different atoms joined by chemical bonds.

Example:

(a) poly silicon dioxide 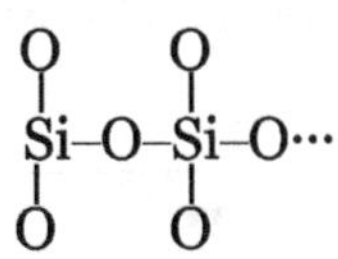

(b) Polyphosphoric acid 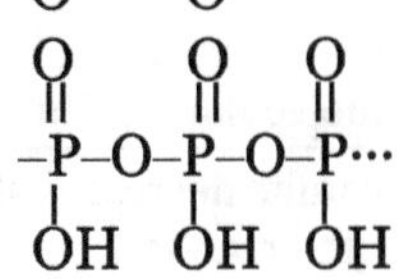

Polymerization

Polymerization is the process of converting small organic molecules into high molecular weight molecules either by addition or by condensation reaction. The small molecules are called monomers and the products are called polymers.

Degree of Polymerization

The number of repeating units present in the polymer chain is called degree of polymerization. Degree of polymerization is given by,

$$\text{Degree of polymerisation } (n) = \frac{\text{Molecular weight of the polymeric network}}{\text{Molecular weight of the repeating unit}}$$

Example:

$$5CH_2 = ch_2 \rightarrow - ch_2 - ch_2 - ch_2 - ch_2 - ch_2 - ch_2 - ch_2 - ch_2 - ch_2 - ch_2$$

Here, five repeating units are present in the polymer chain, so, the degree of polymerization = 5.

Types of Polymerization

There are two types of polymerization:

- Addition polymerization.
- Condensation polymerization.

Polymerization can proceed according to two different mechanisms referred to as:

- Chain growth polymerization.
- Step growth polymerization.

Chain growth polymerization is also called as addition polymerization. This reaction occurs by successive addition of monomer molecules to the reactive end of a growing polymer chain.

In step growth polymerization, reaction between simple polar groups containing monomers with the formation of polymer and elimination of small molecules like H_2O, HCL, etc., occurs.

Addition/chain polymerization	Condensation/step polymerization
Longer reaction time gives higher yield, but have a little effect on molecular weight.	Longer reaction times are essential to obtain high molecular weight.
Number of monomeric units decreases steadily throughout the reaction.	Monomers disappear at the early stage of reaction.
High molecular weight polymer is formed once.	Molecular weight of the polymer rise steadily throughout the reaction.
Molecular weight of the polymer is an integral multiple of molecular weight of monomer.	Molecular weight of the polymer need not be an integral multiple of monomer.

The monomer must have at least one multiple bond. Eg: Ethylene $CH_2 = CH_2$	The monomer must have at least two identical co-different functional group. Eg: Glycol
Monomers add on to give a polymer and no other byproduct is formed.	Monomers condense to give a polymer and byproducts such as H2O, CH3OH are formed.
Homo chain polymer is obtained.	Hetero-chain polymer is obtained.
Thermoplastics are produced. Eg: PVC.	Thermosetting plastics are produced. Eg: Urea - formaldehyde

Additional Polymerization and the Mechanism of Free Radical Polymerization

It is a reaction that yields a polymer, which is an exact multiple of the original monomeric molecule usually containing one or more double bonds. In addition polymerization, there is no elimination of any molecule like HCL, H_2O, etc.

Example:

PVC is produced from vinyl chloride.

$$nCH_2 = \underset{\displaystyle CI}{\underset{|}{CH}} \xrightarrow{\text{Heat / Pressure}} n...(-CH_2-\underset{\displaystyle CI}{\underset{|}{CH}}-)n$$

The mechanism of addition polymerization can be explained by one of the following types:

- Free radical mechanism
- Ionic mechanism

All the above mechanism occurs in three major steps namely:

- Initiation
- Propagation
- Termination

1. Initiation

It is considered to involve two reactions.

i. First reaction: Involves production of free radicals by homolytic dissociation of an initiator or catalyst to yield a pair of free radicals (R*).

$$\underset{\text{(Initiator)}}{I} \text{-------------}> \underset{\text{(Free radicals)}}{2R^*}$$

Commonly used Thermal Initiators

Thermal initiator is a substance used to produce free radicals by homolytic dissociation at high temperature.

$$\underset{\text{Acetyl peroxide}}{CH_3COO-OOCCH_3} \xrightarrow{70-90°C} \underset{\text{Free radicals}}{2(CH_3COO)(or)2R°}$$

ii. Second reaction: Involves addition of this free radical to the first monomer to produce chain initiating species.

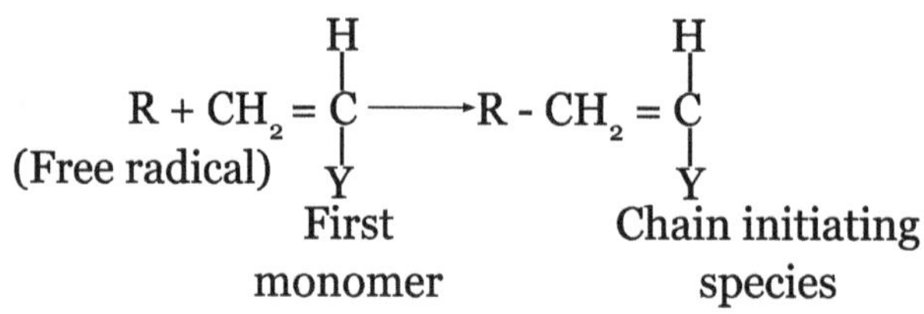

2. Propagation

It involves the growth of chain initiation species by successive addition of large number of monomers.

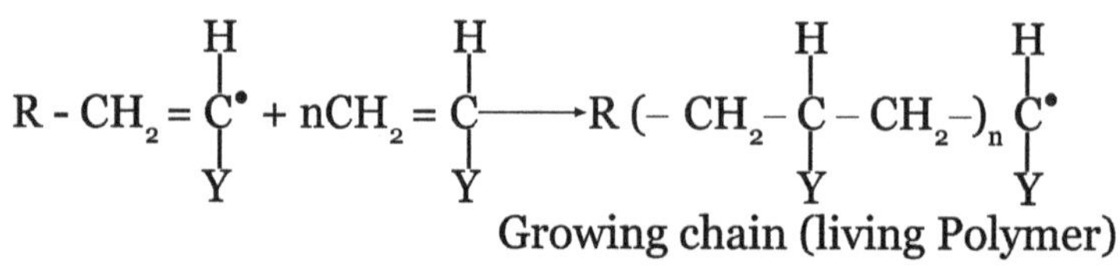

The growing chain of the polymer is known as living polymer.

3. Termination

Termination of the growing chain of the polymer may occur either by:

- Coupling reaction
- Disproportionation

i. Coupling or Combination

It involves coupling of free radicals of one chain end to another free radical forming macromolecule called as dead polymer.

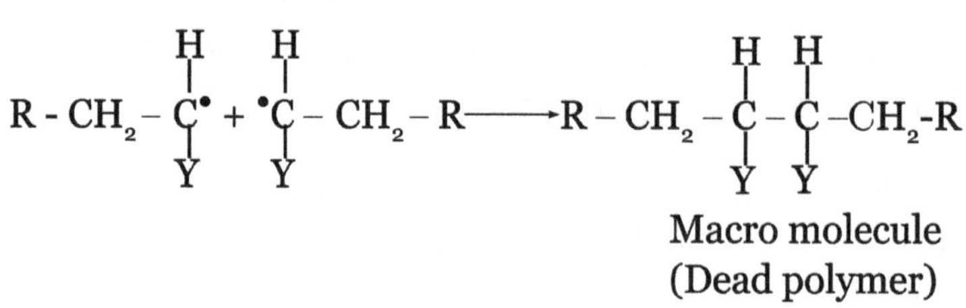

ii. Disproportionation

It involves transfer of hydrogen atom of one radical center to another radical center forming two macromolecules, one saturated and another unsaturated.

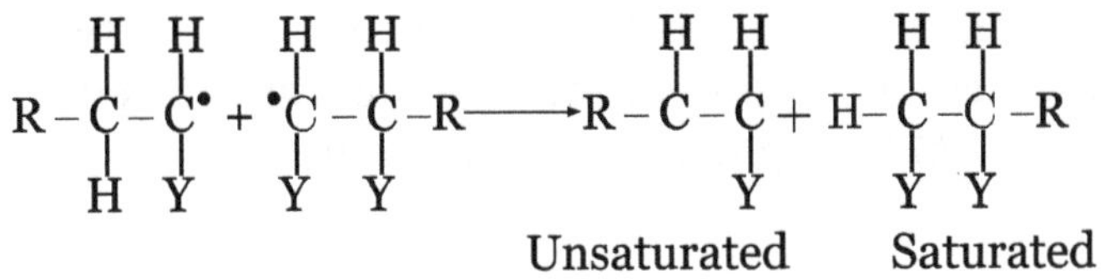

1.1.1 Stereo Regular Polymers

The IUPAC commission on macromolecular nomenclature defines a stereo regular polymer as "a regular polymer whose molecules can be described by only one species of stereo repeating unit in a single sequential arrangement". A stereo repeating unit is further defined as "a configurational repeating unit having defined configuration at all sites of stereo isomerism in the main chain of a polymer molecule".

Tactic polymers are considered to be those which exhibit a regular pattern of configurations for at least one type of stereo isomerism sites in the main chain. According to these nomenclature proposals, one can have tactic stereo regular polymers and tactic non-stereo regular polymers, according to whether every type or only one type of stereo isomerism site in the main chain is in a configurationally ordered sequence. Cases in which steric order exists only for stereo isomerism sites in the side chains, the main chain being non-tactic or atactic are not considered.

1.1.2 Methods of Polymerization (Emulsion and Suspension), Physical and Mechanical Properties

Methods of Polymerization

The following methods are generally used for the polymerization reaction:

- Bulk polymerization.
- Solution polymerization.
- Suspension polymerization.
- Emulsion polymerization.

Bulk Polymerization

Bulk polymerization is the simplest method of polymerization. The monomer is taken in a flask in a liquid form, the initiator and the chain transfer agents are dissolved in it. The flask is placed in thermostat under constant agitation and heated.

Monomer + Initiator + Chain transfer agent → Polymer

The reaction is slow but becomes fast as the temperature increases. After a known period of time, the whole content is poured into methanol and the polymer is precipitated out. PVC, Polystyrene, PMA are prepared by this method.

Advantages:

It is of high purity in the polymer reaction.

Disadvantages:

- Polymerization is highly exothermic.
- During polymerization viscosity of the medium increases, mixing and control of heat is difficult.

Applications:

- Low molecular weight polymers, this method is used in adhesives, plasticizers and lubricant additives.
- Polymer obtained by this method are used in casting formulations.

Emulsion Polymerization

Emulsion polymerization is used for water insoluble monomers and water soluble initiators like potassium persulphate.

Preparation

Monomer + Initiator + Surfactant → Polymer

Advantages:

- The rate of polymerization is high.
- High molecular weight polymer can be obtained.

Disadvantages:

- Polymer needs purification.
- If required rapid agitation can be done.

Applications:

- Emulsion polymerization is used in large scale production.

- In manufacturing tacky polymers like butadiene and chloroprene.

Solution Polymerization

In solution polymerization, the monomer, initiator and the chain transfer agents are taken in a flask and dissolved in an inert solvent. The whole mixture is kept under constant agitation. After some time, the polymer produced is precipitated by pouring it in a suitable non-solvent.

Monomer + Initiator + Chain transfer agent → Polymer

The solvent helps to control heat and reduces viscosity built up. Polyacrylic acid, polyisobutylene are prepared by this method.

Advantages:

- Viscosity built up is negligible.
- The mixture can be agitated easily.

Disadvantages:

- This polymerization requires solvent recovery and recycling.
- The removal of last traces of solvent is difficult.
- It is difficult to get very high molecular weight polymer.

Applications:

If polymer is in liquid form it can be directly used as adhesives and coatings.

Suspension Polymerization

Suspension polymerization is used only for water insoluble monomers. This polymerization reaction is carried out in heterogeneous system. At the end of polymerization, polymer is separated out as pearls or spherical beads. This method is also called pearl polymerization.

Monomer + Initiator + Suspending agent → Polymer

Examples: Polystyrene

Advantages:

- Products obtained are highly pure.
- Isolation of product is very easy.
- Efficient heat control.

- Since water is used as a solvent, this method is more economical.

Disadvantages:

Control of particle size is difficult.

Applications:

Polystyrene beads are used as ion exchangers.

Properties of Polymers

The following are the important properties considered for the polymers:

- Glass transition temperature (T_g).
- Weight average molecular mass $(\overline{M}w)$.
- Tacticity.
- Number average.
- Polydispersity index.

Physical Properties of Polymers:

1. Strength

It depends on the magnitude of force of attraction between polymeric chains. Two types are:

- Primary or chemical bond.
- Secondary or intermolecular forces.

2. Deformation

Deformation is the slipping of one chain over the other or stretching and recoverance of original shape of the polymeric chains.

Types:

- Plastic deformation.
- Elastic deformation.

3. Chemical Resistance and Solubility

Chemical attack is internal, causing softening, swelling and loss of strength of polymer.

The chemical nature of monomeric units and their molecular arrangement determines the chemical resistance of the polymer.

Polymers have polar groups such as -OH, -COOH or Cl dissolve in polar solvents whereas, polymers having non-polar groups $-C_6H_5$ or $-CH_3$ dissolve in non-polar solvents. Polymers of more aliphatic characters are more soluble in aliphatic solvents whereas, polymers of more aromatic character are more soluble in aromatic solvents.

The tendency to solubility or swell of polymers decreases with the increase in molecular weight or chain length of polymer. High molecular weight polymers on dissolving yields solutions of high viscosities.

4. Physical State of Polymers

An amorphous state is characterized by a completely random, irregular and dissymmetrical arrangement of polymeric chains.

Example: Thermosetting, rubber.

A crystalline state is characterized by a completely regular, symmetrical and ordered arrangement of polymeric chains with uniaxial orientation.

Example: Fibers.

A semi-crystalline state consists of crystalline region called as crystallites embedded in an amorphous matrix.

Example: Thermoplastic.

5. Effect of Heat (Glass Transition Temperature T_g)

Mechanical properties of polymers:

i. Hardness

The ability of a polymer to resist scratching, cutting, penetration or abrasion is known as hardness. It is measured by its ability to absorb energy under impact loads.

Hardness is associated with strength. It is closely associated with composition, material structure and other mechanical properties.

ii. Toughness

It is the amount of energy a polymer can absorb before actual fracture or failure takes place. It is the ability of a polymer to withstand shock, vibrations, plastic and elastic deformation.

It is related to impact strength which has the resistance to breakage under high velocity impact conditions, (i.e.), resistance to shock loading.

iii. Stiffness

The resistance of a polymer to elastic deformation, (i.e.) a polymer which suffers slight deformation under load has a high degree of stiffness. Flexibility is the opposite of stiffness.

iv. Density

Relative density is the mass of the polymer with the mass of equal volume of a specific substance. Density is frequently measured as a quality control parameter.

A specimen with smooth surfaces from the crevices and the dust is weighed in air (W_1) and then in freshly boiled water (W_2). It can be written as,

$$\rho_{polymer} = \frac{W_1}{W_1 - W_2}\rho_{water}$$

v. Tensile Strength

The strength of a polymer is its capacity to withstand destruction under the action of loads. It determines the ability of a polymer to withstand stress without failure.

The tensile strength or ultimate strength is the stress corresponding to the maximum load reached before rupturing the polymer.

$$\text{Tensile strength or Stress} = \frac{\text{Force or Maximum Load}}{\text{Area of Cross section}}$$

vi. Abrasion Resistance

It is defined as the ability of a polymer to withstand mechanical action such as scrapping, rubbing and erosion that tends progressively to remove material from its surface. Abrasion is closely related to frictional force, the load and true area of contact, an increase in any one of the three results in greater abrasion or wear. Abrasion process also creates oxidation on the surface from the buildup of localized high temperatures.

vii. Resilience

It is the capacity of a polymer to absorb energy elastically. Resilience is the capacity of a polymer to bear shocks and vibrations. When a body is loaded, it changes its dimension and on the removal of the load it regains its original dimensions.

In fact, the polymer behaves perfectly like a spring, as long as it remains loaded it has stored energy in itself, on the removal of load, the energy stored is given off exactly as in a spring when the load is removed.

viii. Wear and Tear

It occurs when there is a steady rate of increase in the use of polymers in bearing applications and in situations where there is sliding contact. Examples of these are gears, piston rings, seals, cams, etc. Wear and tear is characterized by the fine particles of polymer being removed from the surface or the polymer becomes overheated to the extent where large troughs of melted polymer are removed.

The wear and tear of polymers is extremely complex which depends on the nature of the application and the properties of the material. It is characterized by adhesion and deformation which results in frictional forces that are not proportional to load but rather to speed. The mechanism of wear and tear is complex, the relative rates may change depending on specific circumstance.

1.2 Plastics as Engineering Materials

Plastics

Plastics are products of polymers. Polymers are resins which could be moulded into different shapes by using heat and pressure.

Engineering Materials

Engineering plastics are those which possess physical properties enabling them to perform for prolonged use in structural applications, over a wide temperature range, under mechanical stress and in difficult chemical and physical environments.

Engineering plastics are most frequently thought of as the acetals, nylons, fluorocarbons, phenolics, polycarbonate and poly phenylene oxide. These are indeed engineering materials and are usually used in small amounts, in comparison with the non-engineering plastics which are used in commodity quantities.

Advantages and Limitations:

For most engineering applications, plastics are considered to be competitive primarily with metals, although there is, of course, considerable competition among the plastics themselves. In comparison with metals, plastics have certain properties which are generally considered to be advantageous for engineering applications.

For the most part, plastics have better chemical and moisture resistance and are more resistant to shock and vibration than metals. They are lighter in weight and usually either transparent or at least translucent in thin sections.

They have the advantage of absorbing sound and vibration and some possess greater wear and abrasion resistance than metals. Some of them, as in the case of nylons, for example, are self-lubricating.

Significantly, one of the most important characteristics of plastics is that they usually are easier to fabricate than metals. Some plastics can be plated but perhaps an even more important property is that plastics can be pigmented in a wide variety of colors.

Finally, because of their lighter weight, giving many of them an advantage in cost per cubic inch with respect to metals and because they are usually easier to fabricate, finished parts made of plastics are frequently less costly than those made of metal.

In continuing this comparison of plastics with metals, we must recognize that plastics are characterized also by certain negative properties. Plastics are not as strong as metals. Generally, they possess lower heat resistance and most of them are flammable.

They are characterized by a much larger thermal expansion, are often less ductile than metals and are subject to embrittlement at low working temperatures. Unfortunately, for many applications for which plastics are used or desired to be used, they are more susceptible to creep.

Plastics are softer than metals, when they absorb water or solvents, will change in dimension, which is a negative feature when these materials are used for gears and other parts. In addition, most plastics are subject to degradation by ultraviolet light. Finally, most plastics cost more than metals on a per pound basis and some of them cost more than some metals on a per cubic inch basis.

When plastics are compared with metals on the usual advantage vs. disadvantage consideration, some of their properties can be considered either favourable or unfavourable, depending upon the application. Among these are the softness and flexibility characteristics of plastics, even considering the more rigid types.

Unfilled plastics are thermal and electrical non-conductors and they are subject to deformation by heat or pressure. Fortunately, however, for the engineer and the plastics manufacturer, plastics can be modified by the addition of other materials. In addition, fire retardants, ultraviolet absorbers and other additives can be incorporated into plastics to improve their properties and fillers and reinforcing materials can be incorporated into the plastics to impart many advantages.

Thermoplastic Plastics

- A thermoplastic is a material which becomes soft when heated and hard when cooled. Thermoplastic materials can be cooled and heated several times. They gain or lose hardness with rise or fall in temperature. Their chemical nature does not get affected even on repeated heating and cooling.

- If these resins are softened, they retain their softness at that temperature. These resins can be reclaimed from waste and they are soft, weak and less brittle.
- The method usually used to manufacture polymers is addition polymerization.
- They are generally long chain linear polymers with occasional or no cross linking.
- They are soluble in specific organic solvents.
- Examples of this class of resins are cellulose nitrate, polyacrylates, ethyl cellulose, polyvinyl resins, styrene or polystyrene resins, polyamides, polyethers, polypropylene, polyethylene, etc.

Thermosetting Plastics

- Thermosetting plastics cannot be remoulded because all the polymer chains are linked by strong covalent bond. These are the polymers, which on heating change irreversibly into hard and rigid materials. The melt of this polymer when set into a mould to form an article, is almost a permanent set.
- On reheating, the article does not soften again thus, inhibiting irreversibility. They are hence known as thermo hardening plastics or permanent setting resins and during moulding they acquire three dimensional cross linked structure with strong covalent bonds.
- On reheating, these bonds retain their strength and hence such a plastic does not soften on reheating. Thermo setting plastics cannot be reclaimed from waste due to their irreversibility. They are hard, strong and brittle than thermoplastics.
- The method by which these are formed is called as condensation polymerization.
- They are insoluble in almost all organic solvents, due to their cross linked three dimensional structure.
- Phenol formaldehyde/bakelite, amino plastics, alkyl plastics, epoxy plastics, silicon plastics, etc., are the best known examples of thermosets.

Compounding and Fabrication (4/5 Techniques)

Fabrication of Plastics

Plastic fabrication is the design, manufacture or assembly of the plastic products through any one of the various methods that are present. Some manufacturers prefer fabrication of plastic over working with other materials due to the process's advantages in certain applications. Plastic's malleability and cost effectiveness could make it a versatile and durable material for a range of different products.

The primary processing of plastics involves heating the material until it is liquid or semi-solid, forming it into the required shape and then solidifying it in any one of the following methods:

- By chemical curing in the case of thermosets.
- By simple cooling in the case of thermoplastics.

The fabrication and processing of plastics is highly complex involving many factors such as chemical, mechanical and physical playing an important role.

Coal, petroleum, cotton, wood, gas, air, salt and water are the basic raw materials, binders, fillers, plasticizers, colors, catalysts and lubricants are also added to plastics to provide them with different properties and to enhance their practical applicability.

Compounding or Blending

Compounding is a type of fabrication which can combine two or more plastics into an amalgam, before forming them into a single part. It involves mixing molten plastics to exact specifications and then forming them with a mould, die or other shaping tool.

Compounding is generally used to improve the ease of processing a given material or to enhance the product performance. By combining the advantages and disadvantages of several types of plastic, the process can result in a unique material complementary to a specific application.

Common types of plastic compounds include:

- Base resins.
- Blowing agents.
- Flame retardants.
- Pigment master batches.
- Polymer fillers.
- Purge compounds.

1.3 Preparation, Properties and Applications of Polyethene

Polyethene

The liquid gas under high pressure is pumped into a heated pressure vessel maintained at 15° to 25°C. By the catalytic effect of traces of oxygen present, ethylene is polymerized into poly ethylene. Ethylene-Ethylene bond yields polyethylene $(CH_2CH_2)_n$.

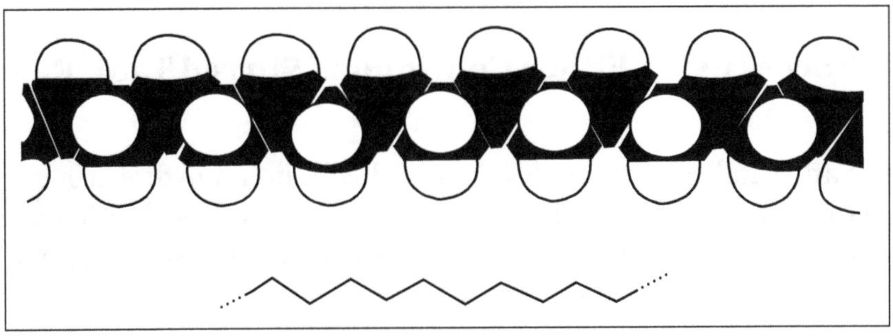

Structure of polyethylene.

Preparation

Polyethene is made by several methods by addition polymerization of ethene, which is principally produced by the cracking of ethane, propane, naphtha and gas oil.

Low Density Polyethene

The process is operated under very high pressure at moderate temperatures as may be predicted from the reaction equation.

$$nC_2H_4 \rightarrow [CH_2 - CH_2]_n$$

$$\Delta H^\ominus = -92 \text{ KJ mol}^{-1}$$

This is a radical polymerization process and an initiator, such as a small amount of oxygen and organic peroxide is used.

Ethene is compressed and passed into a reactor together with the initiator. The molten polyethene is removed, extruded and cut into granules. Unreacted ethene is recycled. The average polymer molecule contains 4000 to 40,000 carbon atoms with many short branches.

High Density Polyethene

Two types of catalyst are primarily used in the manufacture of HDPE. A Ziegler-Natta organometallic catalyst and an inorganic compound, known as a Phillips type catalyst.

A well-known example is chromium (VI) oxide on silica, which is prepared by roasting chromium(III) compound at 1000 K in oxygen and then storing prior to use under nitrogen.

HDPE is produced by three types of process. All operate at relatively low pressure in the presence of a Ziegler-Natta or inorganic catalyst. Typical temperatures range between 350 to 420 K. In all three processes, hydrogen is mixed with ethene to control the chain length of the polymer.

1. Slurry Process using Either Continuous Stirred Tank Reactor or a Loop

The Ziegler-Natta catalyst, as granules, is mixed with liquid hydrocarbon, which simply acts as a diluent. A mixture of hydrogen and ethene is passed under pressure into the slurry and ethene is polymerized to HDPE. The reaction takes place in a large loop reactor with the mixture constantly stirred.

On opening a valve, the product is released and the solvent is evaporated to leave the polymer, still containing the catalyst. Water vapour, on flowing with nitrogen through the polymer, reacts with the catalytic sites, destroying their activity. The residue of the catalyst, titanium(IV) and aluminium oxides remains mixed in minute amounts in the polymer.

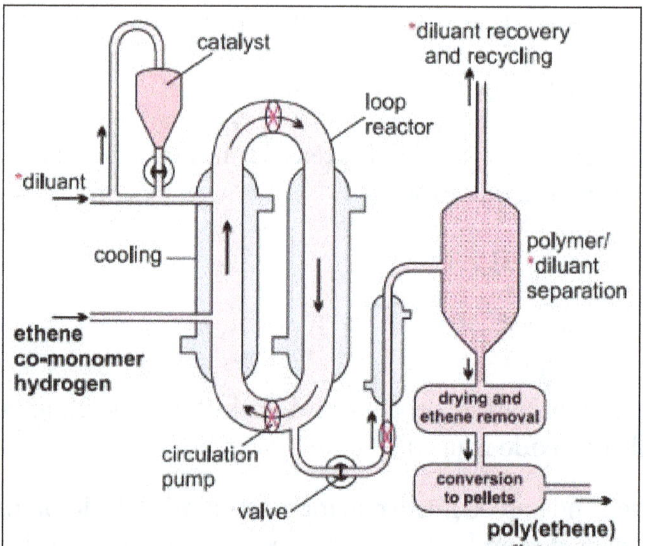

The manufacture of polyethene using the slurry process in a loop reactor.

2. Solution process

The second method involves passing hydrogen and ethene under pressure into a solution of the Ziegler-Natta catalyst in a hydrocarbon. The polymer is obtained in a similar way to the slurry method.

3. Gas phase process

A mixture of hydrogen and ethene is passed over a Phillips catalyst in a fixed bed reactor. Ethene polymerizes to form grains of HDPE, suspended in the flowing gas, which pass out of the reactor when the valve is released.

Modern plants use two or more of the individual reactors in series each of which are under slightly different conditions, so that the properties of different products from the reactors are present in the resulting polymer mixture, leading to a bimodal or broad molecular mass distribution. This provides improved mechanical properties such as stiffness and toughness.

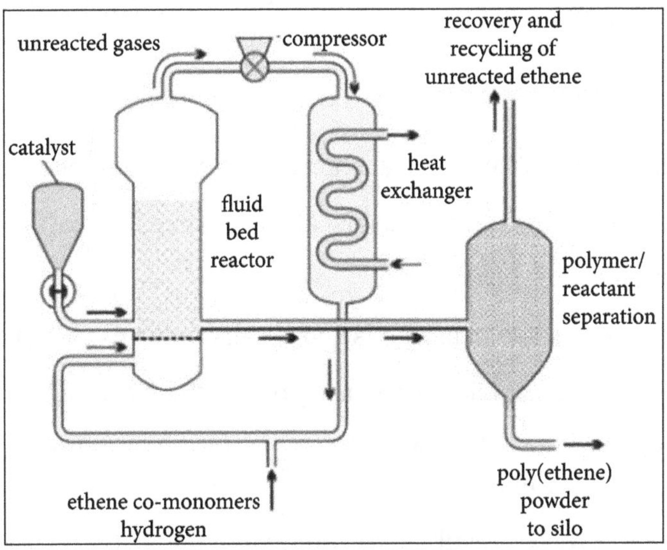

Low pressure gas phase process.

Linear Low Density Polyethene

Low density polyethene has many uses but the high pressure method by which it is produced has high capital costs. However, an elegant technique has been developed, based on both Ziegler-Natta and inorganic catalysts to produce linear low density polyethene LLDPE, which has even improved properties over LDPE. Any of the three processes such as slurry, solution and gas phase can be used when a Ziegler-Natta catalyst is chosen. The gas phase process is used when inorganic catalyst is employed.

Small amounts of a co-monomer such as hex-1-ene or but-1-ene are added to the feedstock. The monomers are randomly polymerized and there are small branches made up of few carbon atoms along the linear chains.

Metallocene Linear Low Density Polyethene

This polyethene, known as mLLDPE, is produced by a new family of catalysts, the metallocenes. Another name for this family is single site catalyst. The benefit is that, mLLDPE is much more homogeneous in terms of molecular structure than classical LLDPE produced by Ziegler-Natta catalysts. Each catalyst is a single site catalyst which produces the same PE chain. There is a transition metal 'filling' a hole between layers of organic compounds.

The catalysts are even more specific than the original Ziegler-Natta and it is possible to control the polymer's molecular mass as well as its configuration. Either the slurry or the solution processes is generally used.

Polyethene produced using a metallocene can be used as a very thin film which has excellent sealing performance and optical properties thus, making them very effective for wrapping foods. The metallocene catalysts have enhanced mechanical properties for the films made out of mLLDPE.

Properties

- It is a rigid waxy solid white, translucent non-polar material.
- Chemically resistant to strong acids, alkalies and salt solutions.
- Good insulator of electricity.
- Swollen and permeable to most oils and organic solvents particularly to kerosene.
- Due to its high symmetrical structure polyethylene crystallizes very easily.
- Polyethylene produced by high pressure process have a branched structure and therefore flexible and tough.
- Low pressure process results in a completely linear PE having better chemical resistance and high density.

Commercial PE is divided into three types. They are:

- Type I or low density PE (0.91-0.925 g/cm^3).
- Type II or medium density PE (0.925-0.940 g/cm^3).
- Type III or high density PE (0.941-0.965 g/cm^3).

Applications:

- Coated wires and cables.
- For making high frequency insulator parts.
- Flexible bottles and bottle caps.

- Kitchen and domestic appliances.
- Sheets for packing materials.
- Toys.
- Tubes and pipes.

1.3.1 PVC

Polyvinyl Chloride

Polyvinyl chloride which is abbreviated as PVC is a thermoplastic resin. PVC has a unique amorphous structure with polar chlorine atoms in the molecular structure. Having chlorine atoms and the amorphous molecular structure are inseparably related. PVC has completely different features in terms of performance and functions as compared with olefin plastics which have only hydrogen and carbon atoms in their molecular structures.

Preparation

Poly vinyl chloride is prepared by polymerization of monomer vinyl chloride:

- About 80% of polymerization includes suspension polymerization.
- 12% emulsion polymerization and 8% bulk polymerization.

Process

VCM and water are introduced into the reactor and the polymerization initiator, along with other additives. The reaction vessel is pressured tight to contain the VCM. The contents of the reaction vessel are mixed continuously to maintain the suspension and ensure a uniform particle size of the PVC resin. The reaction is exothermic and thus requires cooling. As the volume is reduced during the reaction, water is added continuously to the mixture to maintain the suspension.

The polymerization of VCM is started by compounds known as initiators that are mixed into the droplets. These compounds breakdown to start the radical chain reaction. Typical initiators include dioctanoyl peroxide and diacetyl peroxydicarbonate, both of which have fragile O-O bonds. Some initiators initiate the reaction rapidly but decay quickly and other initiators have the opposite effect.

A combination of two different initiators is generally used to give a uniform rate of polymerization. After the polymer has grown by about 10x, the shorter polymer precipitates inside the droplet of VCM and polymerization continues with the precipitated, solvent-swollen particles.

Once it has stopped reacting, the resulting PVC slurry is degassed and stripped to remove excess VCM, which is recycled. The polymer is then passed though a centrifuge to remove water. The slurry is further dried in a hot air bed and the resulting powder is sieved before pelletization or storage. Normally, the resulting PVC has a VCM content of less than 1 part per million.

Properties of PVC

- Weathering stability: PVC is resistant to the aggressive environmental factors therefore; it is a material of choice for roofing.
- Versatility: PVC can either be flexible or rigid.
- Fire protection: PVC is a material that is resistant to ignition due to its chlorine content.
- Longevity: PVC products can last for more than 100 years.
- Hygiene: PVC is one of the material of choice for medical applications, particularly blood and plasma storage containers.
- Energy recovery: PVC has high thermal power, when utilized in incinerators PVC provides power and heat for homes and industries and all that without having any impact on the environment.
- Barrier properties: PVC can be made impervious to the liquids, vapour and gases.
- Eco-efficiency: Only 43% of the PVC's content comes from oil 57% comes from salt, it therefore contributes to the preservation of highly valuable natural resource.
- Recyclability: PVC is recyclable than many other plastics.
- Public safety: PVC has fallen under unfounded attempts so that today it is one of the best explored materials in the world due to scientific researches carried in order to disprove accusations.
- Economic efficiency: PVC is the cheapest of large tonnage polymers providing many products with the best quality-price ratio.

Applications:

- PVC is mainly used as an insulating material.
- It is used for making table clothes, rain coats, toys, tool handles, radio components, etc.
- It is used for making pipes, hoses, etc.

- It is used for making helmets, refrigerator components, etc.
- It is used in making automobile parts and cycle.
- It is used for sewerage pipes and other pipe applications where vulnerability or cost to corrosion limit the use of metal.
- With the addition of impact modifiers and stabilizers, it has become a popular material for door frames and window.
- By adding plasticizers, it can become flexible enough to be used in the cabling applications as a wire insulator.

1.3.2 Bakelite (Phenol-formaldehyde)

Principle

Phenol formaldehyde resins are condensation polymers and are obtained by condensing phenol with formaldehyde in the presence of an alkaline or acidic catalyst. They were first prepared by Backeland, an American Chemist who gave them the name as Bakelite. These are thermosetting polymers.

Preparation

Bakelite is the commercial name for the polymer obtained by the polymerization of formaldehyde and phenol. This is one of the oldest polymers that were synthesized by man. Phenol is made to react with formaldehyde. The condensation reaction of the two reactants in a controlled acidic or basic medium results in the formation of para and ortho hydroxymethylphenols and their derivatives.

The reaction that takes place is as shown below:

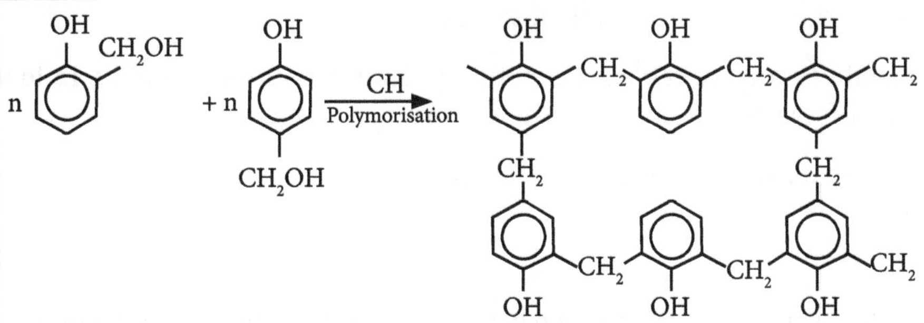

Cross linked polymer - backlite

When the phenol is used in excess and the reaction medium is made acidic, the product of the condensation reaction obtained is novolac. Whereas, when the quantity of formaldehyde taken exceeds the quantity of phenol in the reacting mixture and the reaction occurs in a basic medium, the condensation product obtained is known as resol.

Bakelite has low electrical conductivity and high resistance to heat and chemicals. It is a thermosetting polymer having high strength and capable of retaining its shape after moulding.

Properties

- Phenol formaldehyde resins having low degree of polymerization are generally soft. They possess excellent adhesive properties and are usually used as bonding glue for laminated wooden planks and in varnishes and lacquers.
- Phenol formaldehyde resins having high degree of polymerization are hard, infusible, rigid and scratch resistant. They are resistant to salts, non-oxidizing acids and many organic solvents. They can withstand very high temperatures and act as excellent electrical insulators.
- It does not conduct electricity.
- It is also resistant to chemical action.
- It is resistant to heat and is non-flammable.
- The dielectric constant of bakelite ranges from 4.4 to 5.4.

Applications

- It is used for making bearings, decorative articles, electrical goods, housing laminates, telephone components, TV cabinets, etc.
- Bakelite is used as a substitute of porcelain and other opaque ceramic materials.
- It is used in the area of board and table top games.
- It is used in mounting of metal samples.
- Bakelite is very suitable for emerging electrical and automobile industries because of its extraordinary thermal resistance.
- It is also used as an excellent adhesive.

1.3.3 Teflon

Teflon is a plastic like substance which is produced by polymerizing tetrafluoroethylene $(CF_2 = CF_2)$.

Polytetrafluoroethylene

Polytetrafluoroethylene is also known as teflon. It is a synthetic fluoro polymer of tetrafluoroethylene. It is a polymer which have numerous applications.

The IUPAC (International Union of Pure and Applied Chemistry) name of teflon is Poly (1, 1, 2, 2-tetrafluoroethylene).

Formation of Polytetrafluoroethylene or Teflon or PTFE

It is formed when chloroform is treated with hydrofluoric acid and antimony trifluoride.

$$\underset{\text{tetrafluro ethylene}}{aCF_2 = CF_2} \rightarrow \underset{\text{teflon}}{(CF_2 - CF_2)_n}$$

Properties of Teflon or PTFE

- Teflon is one of the chemically inert substance and it is not affected by strong acids which are chemically harmful and even after by boiling aquaregia.
- It is stable at high temperatures.
- It is a thermoplastic polymer that appears as a white solid at room temperature having a density of about 2200 kg/m^3. It has a melting point of 600 K.
- It bears mechanical properties such that it can degrade gradually at temperatures above 194 K.
- PTFE mainly consist of carbon fluorine bonds and it also gains the properties from the bonds created. Alkali metals and most highly reactive fluorinating agents are the only chemicals that can affect its property.
- It has a co-efficient of friction that is 0.05 to 0.10 which is the third lowest of any known solid material. It has one of the proficient dielectric properties.

Applications of PTFE

- PTFE is usually used to coat non-stick frying pans as it has the capacity to resist high temperatures.
- It is mostly used as a film interface patch for medical and sports applications, having a pressure sensitive adhesive backing. It is installed in one of the high friction areas of footwear, in ankle-foot orthosis, soles, etc.
- It is widely used in medical synthesis, test and many more medicines.

PET

PET is an unreinforced, semi-crystalline thermo-plastic polyester derived from polyethylene terephthalate. Its excellent wear resistance, low co-efficient of friction, high flexural modulus and superior dimensional stability makes it a versatile material for designing electromechanical and mechanical parts. Since PET has no centerline porosity the possibility of leakage and fluid absorption is virtually eliminated.

Products packaged in PET include:

- Baked goods.
- Beauty and household products.
- Beer, wine, spirits and many other food and non-food items.
- Carbonated soft drinks.
- Fresh produce.
- Frozen foods.
- Juice.
- Ketchup.
- Peanut butter.
- Salad dressing.
- Water.

PET container types include:

- Bottles.
- Cups.
- Take-out containers.

1.3.4 Polycarbonates

Polycarbonates are condensation polymers-polyesters of carbonic acid and phenols, HO-CO-OH. They contain [-O-CO-O-] linkage.

Preparation

They are prepared by condensing Bisphenol A with diphenyl carbonate involving elimination of phenol in presence of tertiary amine catalyst.

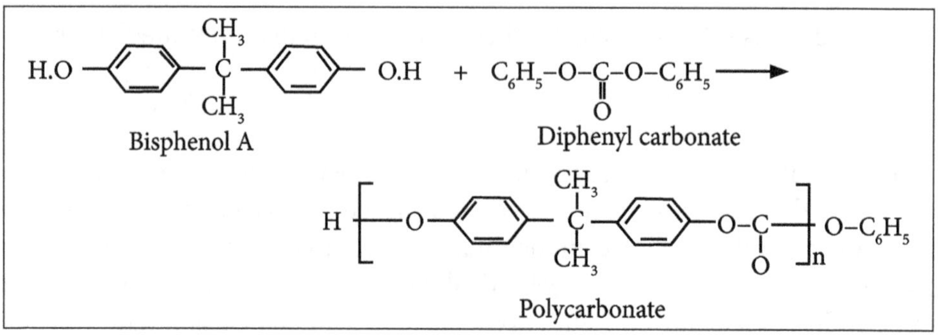

They are also prepared by condensation reaction of sodium salt of Bisphenol A and phosgene.

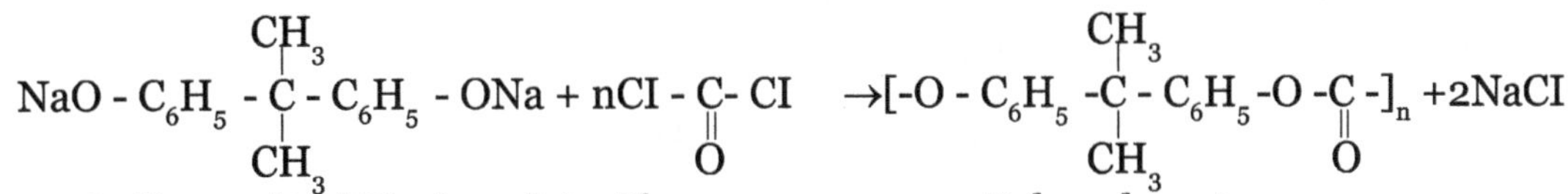

Sodium salt of Bisphenol A Phosgene Polycarbonate

Properties

- Polycarbonate is a transparent thermoplastic.
- It has high melting point, around 265°C.
- It has a high tensile strength and impact resistance.
- It is resistant to water and many organic compounds but alkalies slowly hydrolyze it.

Applications:

- Polycarbonate is widely used due to its unique combination of properties such as extreme toughness, excellent compatibility with several polymers, high heat distortion resistance and outstanding transparency.
- It finds application in making many useful articles such as automobile light lenses, safety goggles, safety shields, telephone parts, etc.
- Polycarbonates are clear thermoplastic polymers which are mainly used as moulding compounds.
- CD-ROMs and baby bottles are well known examples.
- Used as insulator in electronic and electrical industries.

1.4 Elastomers: Natural Rubber

Elastomers or rubbers

Elastomers are linear polymers which have elastic properties.

Depending on the ultimate form and use, polymer can be classified as elastomers, plastic, liquid or fiber resins. When vulcanized into rubbery products exhibiting elongation and good strength, polymers are used as 'elastomers'. Typical examples are synthetic rubber, silicone rubber and natural rubber.

Natural rubber

Rubber is a natural elastic polymer of isoprene. It is obtained from the milk of rubber known as Latex. The structure of natural rubber is as follows:

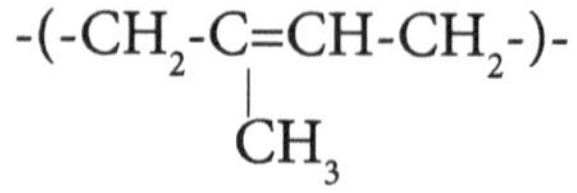

There are plenty of different rubber types available but still the largest single type used is Natural Rubber which is produced from latex from the tree Hevea Brasiliensis. The synthetic types of polymers manufactured from oil have been developed to be used together with Natural Rubber or to replace them or to make polymers with properties superior to Natural Rubber in special areas typically with better outdoor resistance, high temperature resistance or resistance to fuels and oils.

Natural rubber of nearly same characteristics, occurs in the inner bark and to a lesser extent in the leaves and the roots of more than 300 tropical plants such as dandelion, guayule, golden rod, Osage orange. Nowadays 90% of natural rubber is obtained in the form of latex from rubber trees. The latex normally contains 30-65% rubber.

It can be used as such, in the latex form or the solid rubber can be coagulated by the addition of 5% acetic acid or 90% formic acid. Ammonium or potassium alum is also used as coagulants. Natural rubber is an elastic and highly soft material. Natural rubber is a polymerized form of isoprene (2-methyl-1, 3-butadiene):

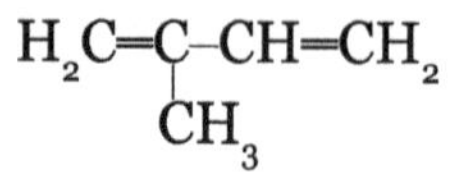

Isoprene is synthesized from propylene:

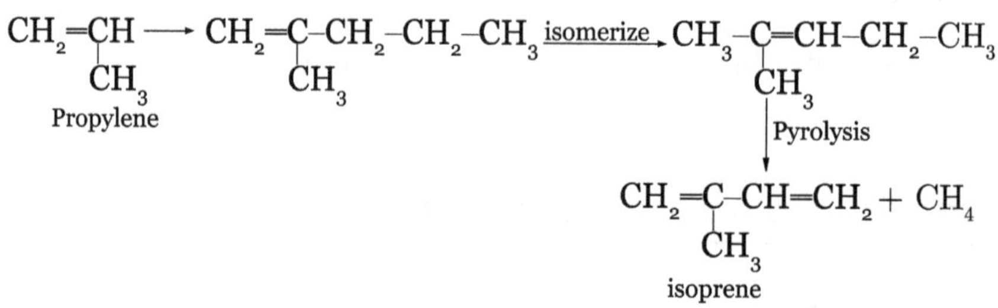

These isoprene molecules polymerize to form, long coiled chains of cis-polyisoprene.

Preparation of Natural Rubber from Latex

- The clear liquid from the top is treated with formic or acetic acid to precipitate rubber.

- Latex is rubber milk containing about 30 to 45% of rubber.
- Rubber sheets are then dried by smoking. This rubber is known as Smoked rubber.
- The rubber milk is diluted with water and allowed to stand for some time.
- The precipitated rubber is collected and passed through rollers to get sheets of rubber.
- During the coagulation of rubber milk with formic or acetic acid, retardants like sodium bisulphite $(NaHSO_3)$ are added to prevent oxidation of rubber. This is known as Creep rubber.

Different Types of Rubber

- General purpose elastomers.
- Special purpose elastomers.
- Speciality elastomers.

General Purpose Elastomers

It comprises of:

- Butadiene rubber (BR).
- Natural rubber (NR).
- Polyisoprene rubber (IR).
- Styrene-butadiene rubber (SBR).

These types have good physical properties, compatibility, good process ability and are generally economical. They are typical polymers used in tyres and mechanical rubber goods with demand for good tensile properties and abrasion resistance. General purpose types constitute the largest volume of polymer used.

Special Purpose Elastomers

It comprises of:

- Acrylonitrile-butadiene rubber or Nitrile rubber (NBR).
- Butyl rubber (IIR).
- Chloroprene rubber (CR).
- Ethylene-propylene rubber (EPM and EPDM).

They all have unique properties which cannot be matched by the general purpose types and are very important for the manufacturing of industrial and automotive rubber products.

The speciality elastomers are polymers with very special properties, in many cases of great importance for the aircraft, automotive, space and offshore industries. Some of these polymers are:

- Acrylic Rubber (ACM).
- Chlorosulphonated Polyethylene (CSM).
- Epichlorohydrine Rubber (CO/ECO/GECO).
- Fluoroelastomers (FPM/FFKM/FEPM).
- Floursilicone Rubber (FPQ).
- Silicone Rubber (PMQ/PV/MQ/VMQ).
- Urethane Rubber (AU/EU).

Disadvantages of Natural Rubber

- Easy ageing.
- Cannot resist strong acid and oil.
- Poor ozone resistance.

1.4.1 Compounding and Vulcanization

Compounding of Rubber

Compounding is mixing of raw rubber with other substances so as to impart the product's specific properties, suitable for a particular job.

The following materials may be incorporated depending on the service conditions of the item to be made from it.

- Softeners and plasticizers.
- Vulcanizing agents.
- Accelerators.
- Antioxidants.
- Colouring matter/pigments.
- Reinforcing fillers.

1. Softeners and Plasticizers

These are added to rubber to offer greater tenacity and adhesion. Important materials are vegetable oils, waxes, stearic acid, etc.

2. Vulcanizing Agents

The main substance added is sulphur. Depending upon the nature of product required, the percentage of "S" added varies between 0.15 to 32%. Many other vulcanizing agents added to rubber are sulphur monochloride, hydrogen sulphide, benzoyl chloride, etc.

3. Accelerators

These materials drastically shorten the time required for vulcanization. Commonly used accelerators are 2-mercaptol, benzthiozol and zinc alkyl xanthate.

4. Antioxidants

Natural rubber undergoes oxidation. For this anti-oxidation, materials such as complex amines and polyphosphites are added, but these materials get darken in the presence of light and therefore, a substance like β-naphthol is used.

5. Colouring Matter/Pigment

These are added to give the rubber product the desired colour. For white products, titanium dioxide pigments are used while for coloured products pigments like chromium oxide → Green, ferric oxide → Red, antimony sulphide → Crimson, lead chromate → Yellow are added.

6. Reinforcing Fillers

These are added to give the strength and rigidity to the rubber products. Most common of all reinforcing fillers especially for the manufacture of motor car tyres is carbon black. Other materials such as ZnO, $CaCO_3$, $MgCO_3$ are widely used.

Vulcanization of Rubber

Vulcanization is compounding of rubber with sulphur. The process is named after vulcan, Roman god of fire. Vulcanization is achieved by heating rubber with sulphur at 140°C in CO_2 atmosphere. Sulphur adds to the double bonds present in rubber to provide cross links between the polymer chains. 2 to 4% sulphur addition gives soft elastic rubber.

When sulphur content is more than 30%, we get hard rubber called 'Ebonite' or vulcanite; this is used to make clarinet mouth pieces and bowling balls.

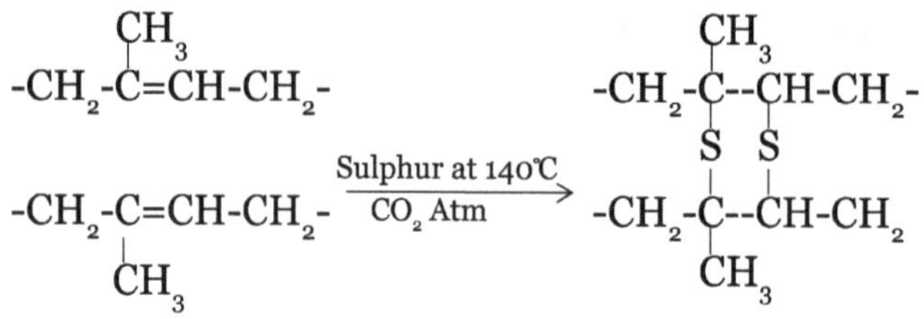

Vulcanization refers to a specific curing process of rubber involving the addition of sulphur and high heat or other equivalent curatives. It is a chemical process in which the polymer molecules are linked to other polymer molecules by atomic bridges composed of carbon to carbon bonds or sulphur atoms.

At the end, the springy rubber molecules becomes cross linked to a greater extent. This makes the bulk material durable, harder and also more resistant to chemical attack. It also makes the surface of the material smoother and prevents it from sticking to plastic or metal chemical catalysts.

This heavily cross linked polymer has strong covalent bonds with strong forces between the chains and therefore, is an infusible and insoluble thermosetting polymer. The vulcanization process is a progressive reaction and is therefore allowed for a specified time. A wide range of products are made with vulcanized rubber including ice hockey pucks, hoses, shoe soles, tires and much more.

Reason for Vulcanizing

Uncured natural rubber is sticky, can easily deform when warm and brittle when cold. In this state, it cannot be used to make articles with a good level of elasticity. The reason for inelastic deformation of unvulcanized rubber can be found in its chemical nature. These polymer chains can move independently relative to each other which results in change of shape.

By the process of vulcanization, cross links are formed between the polymer chains so the chains can no longer move independently. As a result, when stress is applied, the vulcanized rubber will deform but upon the release of stress the rubber article will go back to its original shape.

Properties

- The process of vulcanizing rubber is fairly simple.
- Liquid rubber has sulphur added to it and is then heated at a high temperature.

Elasticity

Vulcanized rubber is just as elastic as it was before the vulcanization process assuming that not more than 10% sulphur was added by weight of the mixture. This adds to the durability of vulcanized rubber enabling it to bend and then move back to its original shape.

Strength

When rubber is vulcanized it becomes cross linked in its chemical structure at the atomic level. This linking of stronger bonds makes vulcanized rubber 10 times stronger than natural rubber. This is one of the vulcanized rubber's greatest strengths as it enables rubber to be used in making more heavy duty products.

Rigidity

Vulcanized rubber is elastic, means it will return to its original shape. As a result of vulcanization, it is also 10 times more rigid than normal rubber.

Uses of Vulcanized Rubber

Vulcanized natural rubber is used for making:

- 5% sulphur is used for making rubber tyre.
- Rubber bands, football bladders, gloves and rubber tubes.
- Tubes and tyres of automobiles and conveyor belts for industrial use after hardening.

1.5 Synthetic Rubbers: Buna S, Buna N, Thiokol and Polyurethanes

Synthetic Rubber

Synthetic rubber is created from petroleum and is classified as an artificial elastomer. This means that it can be deformed without sustaining damage and can return to its original shape after being stretched. Man-made rubber has many advantages over natural rubber and is used in many applications due to its superior performance. In most industrialized nations, the use of synthetic rubber is much more prominent than natural rubber.

Synthetic rubber is used in wide range of applications. In addition to its importance in car tyres, artificial rubber is generally used to produce belts for machinery, medical equipment and moulded parts. Many industrial hoses and seals are also created using man-made rubber.

There are different varieties of synthetic rubber. These are usually created by combining chemicals in different quantities during the rubber production process. Styrene butadiene rubber (SBR) is very common and is able to withstand temperatures between -40 to 212° F. This type of rubber is widely used in conveyor belts, tyre treads for aircraft and automobiles and also for other industrial products.

1. Buna-S or Styrene Butadiene Rubber (SBR)

Buna-S is obtained by co-polymerization of styrene and butadiene in the presence of sodium catalyst. It is also called as styrene rubber or GRS rubber. SBR resembles natural rubber in processing characteristics as well as quality of finished products. It has less tensile strength than natural rubber.

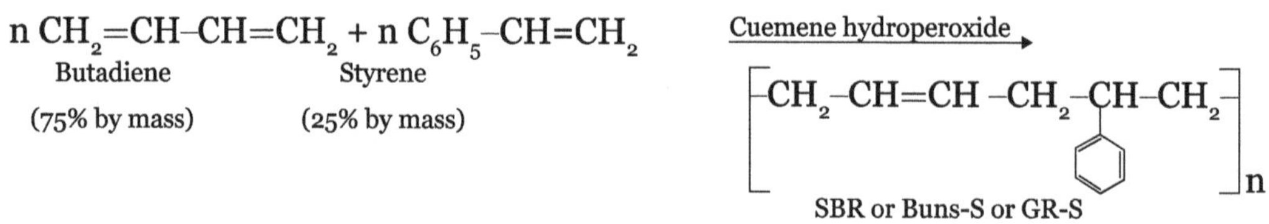

Uses

- It can be vulcanized like natural rubber. It is mainly used in manufacture of tyres.
- It is used as electrical insulator.
- It is used in making floor tiles, gaskets and footwear components, etc.
- It is used in the manufacture of automobile tyres, belts, hoses, rubber soles, etc.

2. Buna-N or Nitrile Rubber

$$\underset{\text{Butadiene}}{n\,CH_2{=}CH{-}CH{=}CH_2} + \underset{\text{Acrylonitrile}}{nCH_2{=}CH{-}CN} \xrightarrow{\text{Copolymerization}} \underset{\text{Nitrile Rubber or NBR}}{\left(-CH_2{-}CH{=}CH{-}CH_2{-}\underset{\substack{|\\CN}}{CH}{-}CH_2{-}\right)_n}$$

It has good tensile strength, abrasion resistance and heat resistance, low swelling and low solubility even after immersion in oils.

Uses

- Nitrile rubbers are used in creamery equipments, fuel tanks, gasoline hoses, etc.
- They also find application in adhesives in the form of latex, for impregnating leather, paper and textiles.

3. Thiokols or Polysulphide Rubber

GR-P is prepared by the reaction between sodium polysulphide (Na_2S) and ethylene dichloride (CH_2Cl-CH_2Cl).

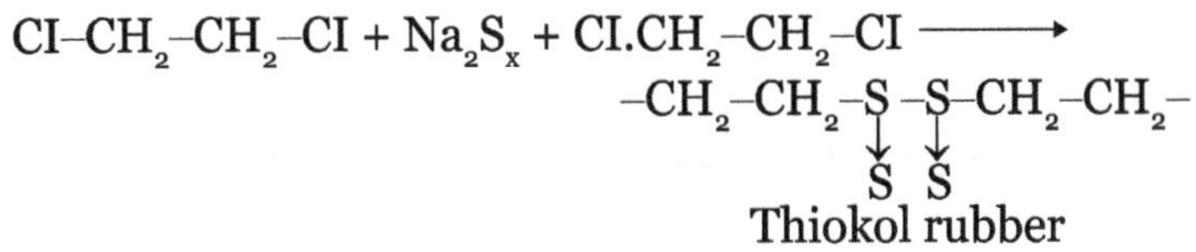

Thiokol rubber possesses extremely good resistance to fuels, mineral oils, oxygen, ozone, solvents and sunlight. It is also impermeable to gases. It cannot be vulcanized. It does not have the properties of rubber. It can withstand cold but not heat.

Uses

- Thiokol rubber are used for barrage balloons, jackets and life rafts which are inflated by CO_2.
- Polysulphide rubbers are also used for lining hoses for conveying gasoline and oil and for printing rolls.

4. Polyurethanes or Isocynate Rubber

Polyurethane rubbers are highly resistant to oxidation and many organic solvents. These are attacked by acids and alkalies. Polyurethane foams are light, tough, abrasion resistant, chemical resistant, heat resistant and weathering resistant.

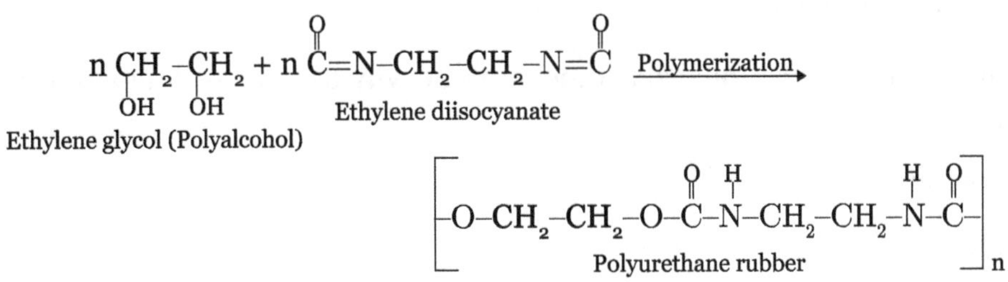

Uses

These rubbers are mainly used in making surface coatings and manufacture of foams and spandex fibers.

1.5.1 Applications of Elastomers

- Construction: Sealants, structural glazing.

- Mould making: Prototyping, restoration, furniture and ornaments.
- Household: Gaskets, O-rings.
- Automotive: Cure-in-place gaskets, vibration dampers.
- Electronics: Adhesives.

Industrial Applications of Elastomeric Materials

1. Gaskets

Gaskets are mechanical seals that fill the space between imperfect mating surfaces. Gaskets are utilized to tightly fill the space. They provide insulation and prevent leakage of liquids or gases between joined surfaces under compressive force.

Most gaskets are customized in size and shape in accordance with the profiles of the joined surfaces. The actual elastomer type to be employed will depend on various factors such as service temperature, environmental properties ranging from weathering, ozone and UV considerations, mechanical properties within the application and to chemical resistance concerns (i.e.) acids, hydraulic fluids, oils, salt water, etc.

Gaskets can be used in many fields. One of the most recognizable one is found in automotive industry, the head gasket. Head gaskets in automotive engines can be made from copper, steel, composite or elastomeric material. This gasket resides between the engine block and the cylinder head in an internal combustion engine. This gasket seals the cylinders to provide optimal compression and prevent leakage of combustible air-fuel mixture, coolant and motor oil.

2. Seals and O-rings

Similar to gasket applications, elastomers are heavily used in sealing applications including:

- Hydraulic
- Pneumatic
- O-ring deployments

Industrial machinery will employ the use of one of these types of seals.

Hydraulic seal applications provide an interesting challenge for elastomeric materials because their function goes just beyond the prevention of fluid leakage. These seals should also withstand extreme temperature ranges, high pressures and transverse forces within the cylinder. The use of hydraulic seals allows fluid power to be converted into linear motion.

Pneumatic seals are typically used in low pressure environments where reciprocating or rotary motions takes place within the cylinder or valve. They can be exposed to high operating speeds where the pressure is not high and requires minimal lubrication. Some examples of pneumatic seals are flange packings, piston seals and rod seals.

The O-ring is a loop of elastomer with rounded cross section that typically fits within the groove and gets compressed between the assemblies of two or more parts. They can be utilized in static or dynamic applications and tends to be popular because they are cost effective, easy to manufacture, simple to assemble and reliable.

3. Noise Reduction and Dampening

The properties of elastomers makes them ideal for many noise reduction and dampening functions. Many elastomeric configurations can reduce resonant vibration and resulting airborne sound. They yield effective vibration damping over a broad frequency and temperature range.

The best example of this application can be found in the automotive industry. Engine mounts and muffler hangers are two prime examples of this use.

Large scale industrial fans also use elastomers to dampen sound. The use of elastomeric fan mounts prevents the transfer of structural vibration from the fan to its mounting structure. Reducing this vibration means less sound radiation and quieter operation.

Examples and Applications of Elastomer Plastic Materials

- Natural rubber: Material used in the manufacture of gaskets, shoe heels.
- Polyurethanes: Polyurethanes are used in textile industry for the manufacture of elastic clothing such as lycra, also used as foam, wheels etc.
- Polybutadiene: Elastomer material used on the wheels or tyres of vehicles, give extraordinary wear resistance.
- Neoprene: Material used primarily in the manufacture of wetsuits is also used as wire insulation, industrial belts etc.
- Silicone: Due to their excellent chemical and thermal resistance, silicones are used in the manufacture of lubricants, medical prostheses, mould, pacifiers, etc.

1.6 Composite Materials and Fiber Reinforced Plastics

Composite Materials

A composite material or composition material or composite is a material made from two or more constituent materials with significantly different physical and chemical properties that, when combined, produce a material with characteristics different from the individual components.

The individual components remains separate and distinct within the finished structure. The new material may be preferred for many reasons, few examples in which materials are stronger, lighter or less expensive when compared to traditional materials.

Fiber Reinforced Plastics

Fiber Reinforced Plastics is the generic term for a uniquely versatile family of composites used from chemical plant to luxury power boats.

Fiber Reinforced Plastic is also known as a fiber reinforced polymer, it is in fact a fiber reinforced plastics composite material constituting a polymer matrix blended with certain reinforcing materials such as fiber. Fibers are generally aramid, basalt, carbon or glass, in certain cases asbestos, paper or wood can also be used.

FRP provides an unrivalled combination of properties:

- Chemical resistance.
- Design freedom.
- Good electrical insulating properties.
- High strength to weight ratio.
- High levels of stiffness.
- Light weight.
- Retention of dimensional stability across a wide range of temperatures.

Applications:

- Building and construction.
- Transportation.
- Marine industry.
- Chemical plant and pipes.

1.6.1 Biodegradable Polymers and Conducting Polymers

Biodegradable Polymers

Polymers are synthetic and natural macromolecules composed of smaller units called monomers. Many synthetic polymers are produced and utilized because they can resist physical and chemical degradation. These polymers which are resistant to degradation results in disposal problems when their usefulness ceases. Research has shown that substitution of natural monomers into synthetic polymers produces polymers that are more easily biodegradable.

A biodegradable polymer is a polymer in which the degradation results from the action of naturally occurring micro-organisms such as algae, bacteria or fungi. These biodegradable polymers are largely used in medical applications where they undergo degradation by chemical hydrolysis.

Biodegradable polymers are a specific type of polymers that breaks down after its intended purpose to result in natural byproducts such as gases CO_2, N_2, water, biomass and inorganic salts.

Polymers prepared from lactic acid and glycolic acid has found a multitude of uses in the medical industry, beginning with the biodegradable sutures first approved in 1960s. Since that time, diverse products based on lactic and glycolic acid and on other materials, including polydioxanone, poly(trimethylene carbonate) co-polymers and polycaprolactone homopolymers, co-polymers have been accepted for use as medical devices. In addition to these approved devices, a great deal of research continues on polyphosphazenes, polyorthoesters, polyanhydrides and other biodegradable polymers.

Classification of Biodegradable Polymers

Biodegradable polymers can be divided into three types:

- Natural polymers.
- Synthetic polymers.
- Modified natural polymers.

Some examples of the biodegradable polymers are shown below.

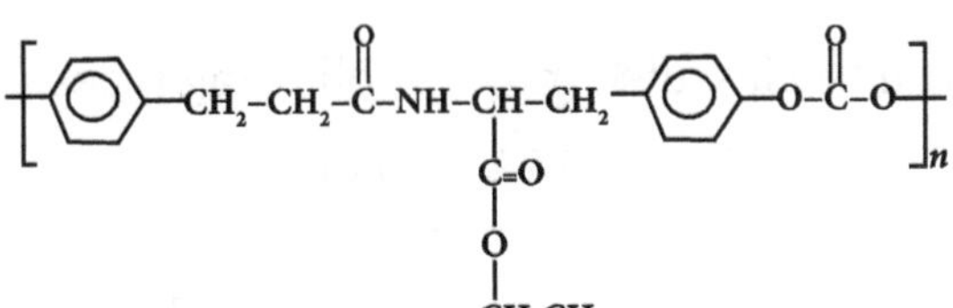

Poly(DTE carbonate)

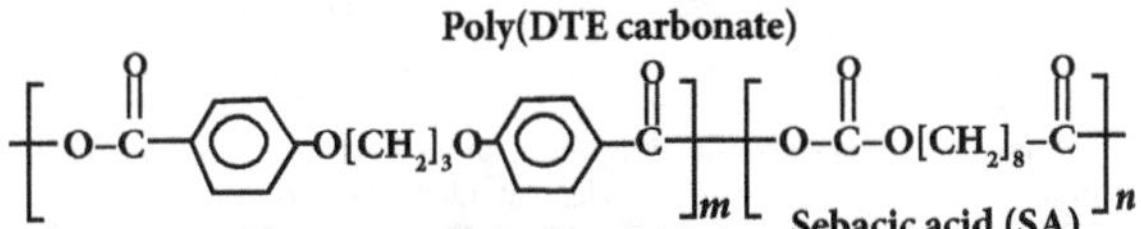

Sebacic acid (SA)

Bis(p-carboxyphenaxy)propane (PCPP)

Poly(PCPP-SA anhydride)

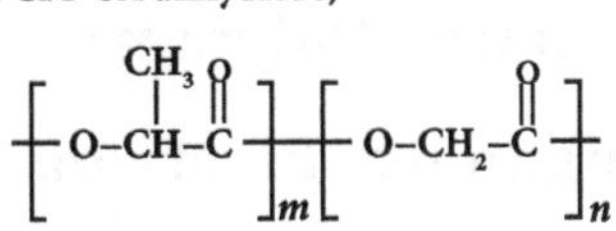

Poly(lactic acid- co-glycolic acid)

Biodegradable polymers.

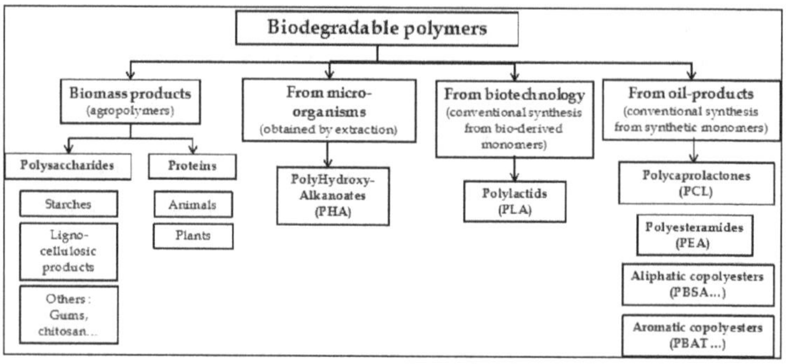

Advantages of Biodegradable Polymers

- Controllable degradation rate.
- Decrease in dosing frequency.
- Improved patient compliance.
- Localized delivery of drug.
- Reduced side effects.
- Sustained delivery of drug.
- Stabilization of drug.

Conducting Polymers

Conducting polymers are long chain having current flowing properties. To make the polymer materials conductive they are doped with atoms that donate positive or negative charges to each unit, enabling current to travel down the chain. Depending on the dopant, conductive polymers exhibit either p-type or n-type conductivity.

Conducting polymers can be classified into the following types.

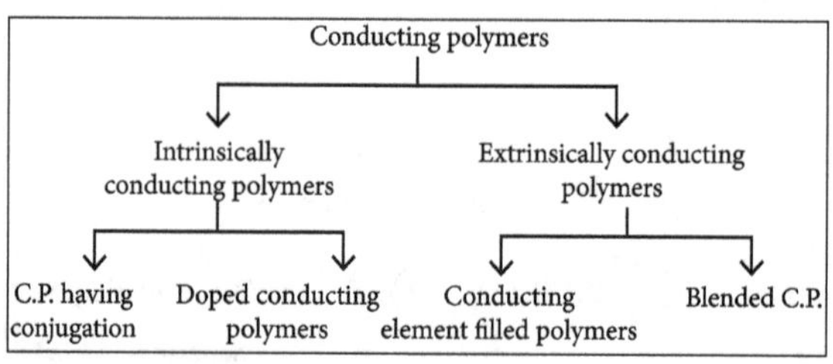

Intrinsically Conducting Polymers

1. CP having Conjugated π-Electrons in the Backbone

Such polymers contains conjugated π-electrons in the back bone which increases their conductivity to a larger extent. This is because, overlapping of conjugated π-electrons over the entire backbone results in the formation of conduction bonds as well as valence bonds, which extends over the entire polymer molecule.

For example, polypyrrole is obtained by electro-polymerization of pyrrole.

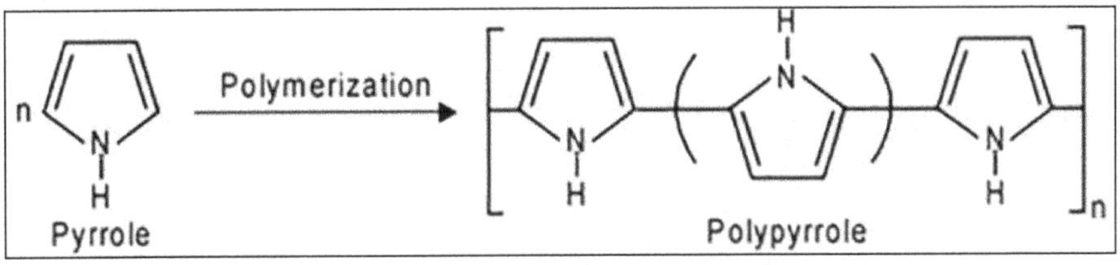

Polypyrrole is an inherently conducting polymer due to interchain dopping of electrons. It can be easily prepared by oxidative polymerization of the monomer pyrrole.

2. Doped Conducting Polymers

It is of two types:

- p-doping (oxidative doping)
- n-doping (reductive doping)

p-doping

It is done by oxidation process. This formation is known as polaron. A second oxidation of this polaron, followed by radical recombination yields two positive charge carriers on each chain which are mobile.

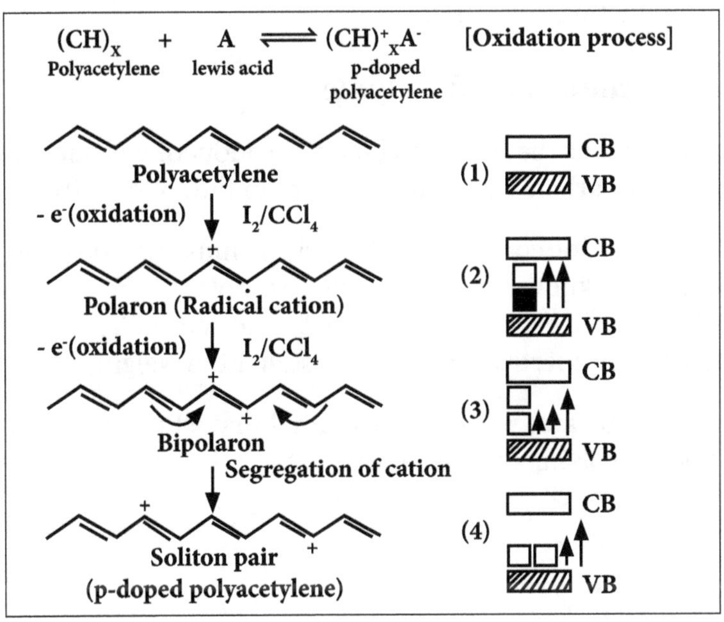

n-doping

It is done by reduction process. It forms polaron and bipolaron in two steps.

This followed by recombination of radicals yields two negative charge carriers on each chain of polyacetylene.

$$\underset{\text{Polyacetylene}}{(CH)_x} + \underset{\text{Lewis acid}}{B} \rightleftharpoons \underset{\text{n-doped polyacetylene}}{(CH)_x^-}\ B^+$$

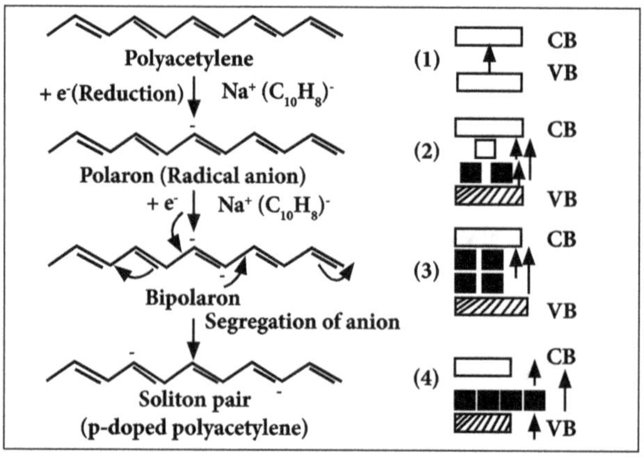

Extrinsically Conducting Polymers

It is of two types:

- Conductive element filled polymers.
- Blended conducting polymers.

1. Conductive Element Filled Polymer

In this, the polymers act as the binder to hold the conducting element such as metallic oxides, metallic fibers, carbon black, etc., together in the solid entity.

Minimum concentration of conductive filler which must be added so that the polymer starts conducting is known as percolation threshold. Carbon black is used as filler which has very high surface area, more porosity and more filamentous properties. It bears good conductive properties and low in cost, light in weight, as well as durable.

It is used in operation theatre where it is essential that, static charges do not build up leading to explosion involving on aesthetics.

2

Fuel Technology

2.1 Fuels: Classification, Calorific Value, HCV and LCV

Fuels

Fuels are materials such as coal, gas or oil that is burned to produce heat or power. The various types of fuels like solid, liquid and gases are available for firing in boilers, furnaces and other combustion equipments.

The selection for right type of fuel depends on various factors such as availability, storage, handling, landed cost of fuel and pollution. The knowledge of the fuel properties helps in selecting the right fuel for the right purpose and efficient use of the fuel.

Classification

Fuels may be broadly classified in two ways, (i.e.):

- According to the physical state in which they exist in nature as solid, liquid and gases.
- According to the mode of their procurement as natural or manufactured.

Some of the fuels are:

Class 1: Wood, paper, cloth, trash and plastics.

- Solid combustible materials that are not metals.

Class 2: Flammable liquids such as gasoline, oil, grease and acetone.

- Any non-metal in a liquid state on fire.

Class 3: Electrical, energized electrical equipment.

- As long as it's "plugged in," it would be considered a class C fire.

Class 4: Metals such as potassium, sodium, aluminum and magnesium.

Calorific Value, Higher and Lower Calorific Values

Calorific Value

Calorific value is a measure of heating power and it depends on the composition of gas. CV refers to the amount of energy released when a known volume of gas is completely combusted under specified conditions.

The calorific value is the measurement of heat or energy produced and it is measured as:

- Gross calorific value.
- Net calorific value.

The difference in latent heat of condensation of the water vapour is produced during the combustion process.

Gross calorific value assumes that, all the vapour which is produced during the combustion process is fully condensed.

Net calorific value assumes that, the water leaves with the combustion products without it is fully being condensed. Fuels should be compared based on the net calorific value.

Thus, the calorific value of coal varies considerably depending upon the ash, moisture content and the type of coal while calorific value of fuel oils are much more consistent.

Higher Calorific Value (HCV)

Higher calorific value of a fuel is defined as the amount of heat evolved when a unit weight or the volume in case of gaseous fuels are completely burnt and the products of combustion are cooled to normal conditions. The heat contained in the water vapour should be recovered in the condensation process.

Lower Calorific Value (LCV)

Lower calorific value of a fuel is defined as the amount of heat evolved when a unit weight or the volume in case of gaseous fuels are completely burnt and the water vapour leaves with the combustion products that are not condensed.

Some typical values	HCV	LCV
Anthracite coal	34.583	33.913 MJ/kg
Wood	15.826	14.319

Petrol	46.892	43.710
Diesel	45.971	43.166
U.K. natural gas	36.38	32.75 MJ/m3 **
Hydrogen	11.92	10.05

** At SSL conditions (15 °C and 101.325 kPa)

Principle

The gross calorific value of a solid or liquid fuel, when burnt in excess air or oxygen is calculated. The heat liberated is absorbed by surrounding water and copper calorimeter.

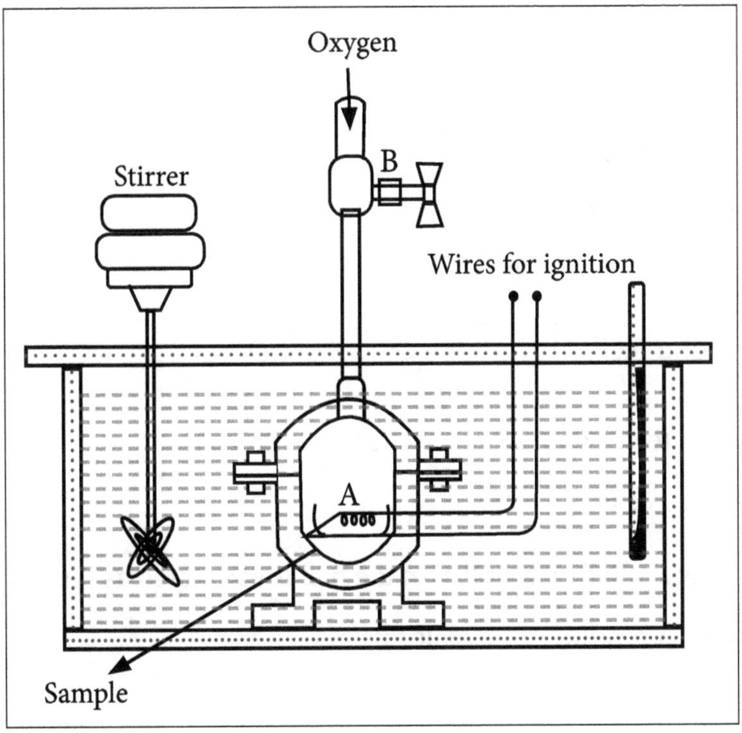

Construction

The bomb calorimeter consists of a cylindrical steel vessel (bomb A) with an airtight screw lid and an inlet valve B, for pumping oxygen. The bomb has a platinum crucible with a loop of wires or electrical ignition coil for initial combustion of fuel. The bomb is kept in a rectangular copper vessel containing weighed amount of water. The equipment is provided with a mechanical stirrer for uniform heat distribution and a Beckmann thermometer to note the temperature. Calorimeter is enclosed in a jacket to minimize the heat exchange with the surroundings.

Working

A known mass of fuel is made into a pellet and taken in the crucible and oxygen is passed through the inlet valve. A known mass of water is taken in the calorimeter and is closed with the lid. The initial temperature of water is noted. The fuel is ignited by passing current through the coil. The heat released due to burning of fuel is absorbed by water. The temperature of water rises. The final temperature of water is noted on the thermometer.

Calculation

Let,

Mass of fuel = m kg

Mass of water = w_1 kg

Water equivalent of calorimeter = w_2 kg

Initial temperature of water = t_1 °C

Final temperature of water = t_2 °C

Specific heat of water (s) = kJ kg^{-1} °C^{-1}

$$\therefore \text{GCV}(\text{solid fuel}) = \frac{(w_1 + w_2) \times (t_2 - t_1) \times S}{m} \text{ kJ kg}^{-1}$$

NCV can be calculated if the % of H_2 in the fuel is known,

$$2\text{ H} + \frac{1}{2}O_2 \rightarrow H_2O$$

If the fuel contains x% hydrogen, NCV of the fuel is calculated as follows.

2 atoms of hydrogen produces one molecule of water.

2g of hydrogen produces 18 g of water.

(i.e.) x g of hydrogen produces 9 x g of water.

x % hydrogen (9/100) x g of water = 0.09 x g of water

NCV = GCV - latent heat of steam formed

= GCV - 0.09 × % of H_2 × latent heat of steam

= GCV - 0.09 × % of H_2 × 587 cal/g (OR)

NCV = GCV - 0.09 × % of H_2 ×2454 kJ kg^{-1}

2.1.1 Dulong's formula and Bomb calorimeter

Dulong's Formula

Dulong suggested a formula for the calculation of calorific value of the solid or liquid fuels from their chemical composition which is as given below:

Gross calorific value,

$$\text{H.H.V.} = \frac{1}{100}\left[33800\ \text{C} + 144000\left(\text{H} - \frac{\text{O}}{8}\right) + 9270\ \text{S}\right]\text{kJ/kg} \quad ...(1)$$

Or,

In MKS units,

$$\text{H.C.V.} = \frac{1}{100}\left[8080\ \text{C} + 34500\left(\text{H} - \frac{\text{O}}{8}\right) + 2240\ \text{S kcal/kg}\right]$$

Where C, H, O and S are carbon, hydrogen, oxygen and sulphur in percentages respectively in 100 kg of fuel. In the above formula, oxygen is assumed to be in combination with hydrogen and only extra surplus hydrogen supplies the necessary heat.

Bomb Calorimeter (Laboratory Method)

The calorific value of solid and liquid fuels is determined in the laboratory by Bomb calorimeter. It is so named because its shape resembles that of a bomb. Figure shows the schematic sketch of a bomb calorimeter.

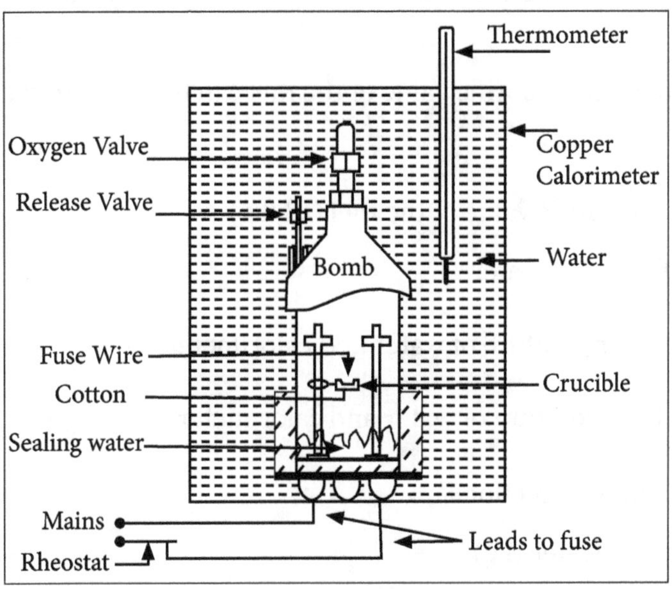

Bomb calorimeter.

The calorimeter is made of austenitic steel which provides considerable resistance to corrosion and allows it to withstand high pressure. In calorimeter, there is a strong cylindrical bomb in which combustion occurs. The bomb has two valves at the top. One supplies oxygen to the bomb and other releases the exhaust gases.

A crucible in which a weighted quantity of fuel sample is burnt is arranged between the two electrodes as shown in figure. The calorimeter is fitted with water jacket, which surrounds the bomb. To reduce the losses due to radiation, calorimeter is further provided with a jacket of air and water. A stirrer for keeping the temperature of the water uniform and a thermometer to measure the temperature up to accuracy of 0.001°C is fitted through the lid of the calorimeter.

Procedure

About 1 gm of fuel sample is accurately weighed in the crucible and a fuse wire is stretched between the electrodes. It should be ensured that the wire is in close contact with the fuel. To absorb the combustion products of nitrogen and sulphur, 2 ml of water is poured in the bomb. The bomb is then supplied with pure oxygen through the valve to an amount of 25 atmosphere. The bomb is then placed in the weighed quantity of water, in the calorimeter.

The stirring is started after making the necessary electrical connections and when the thermometer indicates a steady temperature, fuel is fired and temperature readings are recorded after 1/2 minute intervals until maximum temperature is attained. The bomb is then removed, the pressure is slowly released through the exhaust valve and the contents of the bomb are weighed carefully for further analysis.

The heat released by the fuel on combustion is absorbed by the calorimeter and the surrounding water.

From the above data, the calorific value of the fuel can be found in the following way:

Let,

C = Higher calorific value of the fuel (kJ/kg)

c = Specific heat of water

t_1 = Initial temperature of water and calorimeter

t_2 = Final temperature of water and calorimeter

t_c = Radiation corrections

w = Weight of water (kg)

w_f = Weight of fuel sample (kg)

w_e = Water equivalent of calorimeter (kg)

Heat released by the fuel sample $= w_f \times C$

Heat received by water and calorimeter $= (w_w + w_e) \times c \times [(t_2 - t_1) + t_c]$

Heat lost = Heat gained

$$\therefore w_f \times C = (w + w_e) \times c \times [(t_2 - t_1) + t_c]$$

$$C = \frac{(w + w_e) \times c \times [(t_2 - t_1) + t_c]}{w_f}$$

[Value of c is 4.18 in SI units and unity in MKS units]

2.1.2 Numerical Problems

1. Let us calculate the gross calorific value and net calorific value of a sample of coal. 0.5g of coal which when burnt in a bomb calorimeter raised the temperature of 1000g of water from 293K to 296.4K. The water equivalent of calorimeter is 350 g. The specific heat of water is 4.187 kJ/kg °K, latent heat of steam is $2457.2 \text{ kJ kg}^{-1}$. The coal sample contains 93% carbon, 5% hydrogen and 2% ash.

Solution:

Given:

m = Mass of the fuel = 0.5 g

w_1 = Mass of water taken = 1000 g

w_2 = Water equivalent of calorimeter = 350 g

t_1 = Initial temperature of water = 293 K

t_2 = Final temperature of water = 296.4 K

S = Specific heat of water = 4.187 kJ/kg °K

Formula to be used:

$$GCV(\text{solid fuel}) = \frac{(w_1 + w_2) \times (t_2 - t_1) \times S}{m} \text{ kJ kg}^{-1}$$

NCV (solid fuel) = GCV - latent heat of steam formed.

$$= \text{GCV} - (0.09 \times \% \text{ of } H_2) \times \text{ latent heat of steam}$$

$$= \text{GCV (solid fuel)} = \frac{(w_1 + w_2) \times (t_2 - t_1) \times s}{m} \text{ kJ kg}^{-1}$$

$$= \frac{(1000 + 350)\text{g} \times (296.4 - 293)\text{K} \times 4.187 \text{ kJ kg}^{-1}\text{ K}^{-1}}{0.5 \text{ g}}$$

$$= \frac{1350 \text{ g} \times 3.4 \text{ K} \times 4.187 \text{ kJ kg}^{-1}\text{ K}^{-1}}{0.5 \text{ g}}$$

$$= 38{,}437 \text{ kJ kg}^{-1}$$

NCV (solid fuel) = GCV - latent heat of steam formed

$$= \text{GCV} - (0.09 \times \% \text{ of } H_2) \times \text{ latent heat of steam}$$

$$= 38437 \text{ kJ kg}^{-1} - (0.09 \times 5) \times 2457.2 \text{ kJ kg}^{-1}$$

$$= 38437 \text{ kJ kg}^{-1} - 1106 \text{ kJ kg}^{-1}$$

$$= 37{,}331 \text{ kJ kg}^{-1}$$

2. Let us calculate the gross and net calorific value of a coal sample from the following data obtained from bomb calorimeter experiment.

Solution:

Given:

Weight of coal (m) = 0.73 g

Weight of water taken in calorimeter $(w_1) = 1500$ g

Water equivalent of calorimeter $(w_2) = 470$ g

Initial temperature $(t_1) = 25$ °C

Final temperature $(t_2) = 27.3$ °C

Percentage of hydrogen in coal sample = 2.5%

Latent heat of steam = 587 Cal/g

$$GCV(\text{solid fuel}) = \frac{(w_1 + w_2) \times (t_2 - t_1) \times S}{m} \text{ kJ kg}^{-1}$$

Formula to be used:

NCV (solid fuel) = GCV - latent heat of steam formed

$$= GCV - (0.09 \times \% \text{ of } H_2) \times \text{latent heat of steam}$$

$$GCV(\text{solid fuel}) = \frac{(w_1 + w_2) \times (t_2 - t_1) \times s}{m} \text{ kJ kg}^{-1}$$

$$= \frac{(1500 + 470) \times 10^3 \text{ kg} \times (27.3 - 25)\,^{o}C \times 4.187 \text{ kJ kg}^{-1}\,^{o}C^{-1}}{0.73 \times 10^3 \text{ kg}}$$

$$= 25988 \text{ kJ kg}^{-1}$$

Net calorific value = HCV - Heat released by the condensation of steam

$$= GCV - 0.09 \times \% H_2 \times \text{Latent heat of steam}$$

$$= 25988 - 0.09 \times 2.5 \times 2454$$

$$= 25435.85 \text{ kJ kg}^{-1}$$

2.2 Coal

Coal is a natural fuel formed by the slow carbonization of vegetable matter buried under the earth some thousands of years ago. It is classified into four kinds based on the carbon content and the calorific value.

They are:

- Peat
- Lignite
- Bituminous coal
- Anthracite coal

Peat

It is the first stage of formation of coal from wood. Peat is brown in color and fibrous jelly-like mass. It contains 80-90% moisture. The calorific value of peat is 5400 kcal/kg. The composition of peat is C = 57%, H = 6%, O = 35% ash = 2.5%. It is a low grade fuel due to high water content.

Uses

It is used as a fertilizer and packing material.

Lignite

Lignite is known as immature form of coal. It contains 20-60% moisture. The air dried lignite contains C = 60-70% O = 20%. This burns with a long smoky flame. The calorific value of lignite is 6500-7100 kcal/kg.

Uses

- This is used as a domestic fuel.
- It is used as a boiler fuel for steam rising.
- It is used in the manufacture of producer gas.

Bituminous coal

It is a high quality fuel. Its moisture content is 4%. Its calorific value is 8500 kcal/kg. It's composition is C = 83%, O = 10%, H = 5% and N = 2%.

Uses

- This is used in metallurgy.
- It is used for making coal gas.
- This is used in steam rising.
- It is also used for domestic heating.

Anthracite coal

It is the superior form of coal. It contains C = 92-98%, O = 3%, H = 3% and N = 0.7%. It burns without smoke. It's calorific value is 8700 kcal/kg.

Uses

- This is used for steam raising and house hold purposes.
- It is used in metallurgy and for direct burning in boilers.
- It is used in coal tar distillation.

- It is used in thermal power plant.
- It is used in glass furnaces.

2.2.1 Proximate and Ultimate Analysis, Significance of the Analyses

Proximate Analysis

Proximate analysis includes the determination of moisture, volatile, ash and fixed carbon content.

1. Determination of Moisture Content

1 g of finely powdered air dried sample is taken in a crucible and is heated in an electrically heated hot air oven at a temperature of 105°C - 110°C for 1 hour. After heating, the crucible is taken outside from the air oven and cooled in a desiccator and weighed.

$$\text{Percentage of moisture content} = \frac{\text{Loss in weight of coal}}{\text{Weight of coal take}} \times 100$$

2. Determination of Volatile Matter

The dried sample is left in the crucible along with the lid that is heated in a muffle furnace at a temperature of 950 ± 25°C for 7 minutes and then cooled in a desiccator and weighed.

$$\text{Percentage of volatile content} = \frac{\text{Loss in weight of coal}}{\text{Weight of coal take}} \times 100$$

3. Determination of Ash Content

The residual sample after the above two experiments in the crucible is being heated in the furnace at 750°C for 30 minutes without the lid. Then it is cooled in a desiccator and weighed.

$$\text{Percentage of ash content} = \frac{\text{Weight of ash left}}{\text{Weight of coal take}} \times 100$$

4. Determination of Fixed Carbon

Here, fixed carbon content can be determined indirectly by subtracting the percentage of moisture, volatile and ash content from 100.

% Fixed carbon = 100 - % of (moisture + volatile matter + ash).

Significance

- It has higher percentage of moisture content which lowers the calorific value of coal. Hence lower the moisture content better the quality of coal.
- It has higher percentage of volatile matter which reduces the calorific value of coal. Low volatile matter shall also reduce the coking property of coal.
- The ash being non-combustible, reduces the calorific value of coal. The ash deposition also causes problems in the furnace walls and the ultimate disposal of ash is also a problem.
- If it has higher percentage of fixed carbon, then higher is the calorific value and better is the quality of coal.

Ultimate Analysis

It refers to determination of weight percentage of sulphur, oxygen, nitrogen, hydrogen and carbon.

1. Carbon and Hydrogen

A required amount of coal sample is taken and burnt in a current of O_2 in combustion apparatus whereby CO_2 and H_2O are formed. The CO_2 and H_2O are absorbed by previously weighed tubes containing KOH and anhydrous $CaCl_2$. The increase in weight gives the C and H content as follows.

$$C + O_2 \rightarrow CO_2$$

$$H_2 + \frac{1}{2} O_2 \rightarrow H_2O$$

$$2KOH + CO_2 \rightarrow K_2CO_3 + H_2O$$

$$CaCl_2 + 7H_2O \rightarrow CaCl_2 \cdot 7H_2O$$

$$\text{Percentage of C} = \frac{\text{Increase in wt. KOH} \times 12 \times 100}{\text{Wt. of coal sample} \times 44}$$

$$\text{Percentage of H} = \frac{\text{Increase in wt. } CaCl_2 \times 2 \times 100}{\text{Wt. of coal sample} \times 18}$$

2. Nitrogen

About 1 g of accurately weighed coal sample is taken in a long necked flask along with the concentrated H_2SO_4, K_2SO_4 and heated. Then it is treated with excess of NaOH and the liberated NH_3 is absorbed in known excess of standard acid solution. The excess acid is back titrated with standard NaOH solution. From the volume of acid consumed, the N content is calculated as follows.

$$\% \text{ of N} = \frac{\text{Volume of acid consumed} \times \text{Normality}}{\text{Weight of coal take}} \times 1.4$$

3. Sulphur

While determining the calorific value of coal sample in a bomb calorimeter, the S in the coal is converted to sulphate. Finally, the washings containing sulphate is treated with dilute HCL and $BaCl_2$ solution which precipitates $BaSO_4$ which is filtered in a sintered glass crucible, washed with water and heated to a constant weight.

$$\% \text{ of S} = \frac{\text{Weight of } BaSO_4}{\text{Weight of coal taken}} \times \frac{32}{233} \times 100$$

4. Oxygen

It is determined indirectly by subtracting the combined percentage of carbon, hydrogen, sulphur and nitrogen from 100.

% of oxygen = 100 - % of (C + H + S + N).

Significance

- Higher percentage of carbon and hydrogen increases the calorific value of coal and hence better is the coal.
- Higher the percentage of O_2 lower is the coking power and lower is the calorific value. Also O_2 when combined with hydrogen in the coal, hydrogen available for combustion becomes unavailable.
- S although contributes to calorific value is undesirable, due to its polluting properties as it forms SO_2 on combustion.

Problems

1. The ultimate analysis of a coal sample indicates Carbon = 84%, Sulphur = 1.5%, Nitrogen = 0.6%, Hydrogen = 5.5% and Oxygen = 8.4%. Let us calculate the GCV.

Solution:

Given:

C = 84%

S = 1.5%

$N = 1.6\%$

$H = 5.5\%$

$O = 8.4\%$

Formula to be used:

$$GCV=\frac{1}{100}\left[8080C+34500\left[H-\frac{O}{8}\right]+2240S\right]Kcal/kg$$

$$GCV=\frac{1}{100}\left[8080C+34500\left[H-\frac{O}{8}\right]+2240S\right]Kcal/kg$$

$$=\frac{1}{100}\left[8080\times 84+34500\left[5.5-\frac{8.4}{8}\right]+2240\times 1.5\right]Kcal/kg$$

$$=8356.05\ K.cal/kg.$$

2.3 Liquid Fuels, Petroleum and Refining

Liquid Fuels

Liquid fuels are those which are combustible, energy generating substances and play vital role in transportation and economy. Most widely used liquid fuels are derived from fossil fuel, petroleum and crude oil. Some important liquid fuels are petrol, kerosene, diesel, etc.

Classification of Liquid Fuels

The liquid fuels can be classified as follows:

- Natural or crude oil.
- Artificial or manufactured oils.

Advantages

The advantages of liquid fuels are as follows:

- Liquid fuels occupy less storage space than solid fuels.
- As compared to solid fuels, they have a high calorific value.
- They can be easily transported through pipes.

- During burning, liquid fuels do not yield any ash or residue.
- The burning process of liquid fuels is clear.
- The combustion is uniform and very easily controllable.
- For complete combustion of liquid fuels, less air is required than that of the solid fuels and hence, their thermal efficiency is high.
- They can be used in IC engines, boilers and gas turbines.
- They do not undergo spontaneous combustion.

Disadvantages

The disadvantages of liquid fuels are as follows:

- When the liquid fuels undergo incomplete combustion, they give unpleasant odour.
- In comparison with solid fuels they are costly.
- Risk of fire hazards are more in case of inflammable and volatile liquid fuels. Thus, they should be stored and transported more carefully.
- During storage, some amount of liquid fuels may escape due to evaporation.
- Special type of burners and sprayers are required for effective combustion.

Petroleum

Petroleum is a naturally available liquid fuel. This is a dark greenish brown viscous oil found deep in earth's crust. It is composed of various hydrocarbons with a small amount of other organic compounds as impurities.

Refining of Petroleum

The process of purification and separation of various fractions present in petroleum by fractional distillation is called refining of petroleum. Refining is done in oil refineries. Petroleum is a complex mixture of organic liquids also known as crude oil or fossil fuel. It is formed from the fossilized dead plants and animals by exposure to heat and pressure in the Earth's crust and was formed millions of years ago.

It is a viscous dark coloured, foul-smelling liquid along with water and soil particles. Hence, it is necessary to separate these hydrocarbons into useful products and this process is known as fractional distillation. In this process, products are separated depending on boiling points, known as refining of petroleum. Refining of petroleum involves the following three steps.

Step 1: Separation of Water by Cottrell's Method

Petrol or crude oil is the emulsion of oil and salt water and these colloidal water droplets coalesce to form large drops which can separate out from oil when the crude oil is sent through two highly charged electrodes.

Step 2: Removal of Sulphur Compounds

Crude oil is treated with copper oxide, sulphur reacts with copper to form copper sulphide precipitate, which is removed by filtration.

Doctor Sweetening Process

The process was described by G. Wendt and S. Diggs. Here, crude oil is treated with sodium plumbate, (i.e.) doctors solution, converts mercaptans in sour gasoline into disulphide.

$$\underset{\text{Mercaptan from Gardinc}}{RSH} + \underset{\text{(sodium plubate)}}{NO_2PbO_2} \xrightarrow{\text{powered S in presence of NaOH}} \underset{\text{(alkyl disulphide)}}{R-S-S-R} + PbS + 2NaOH$$

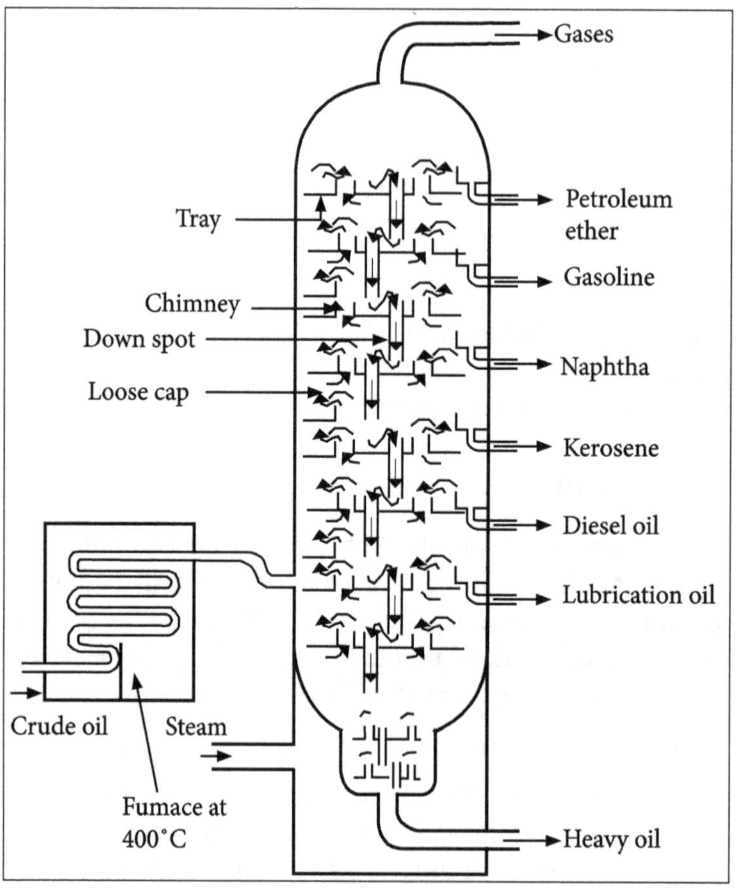

Fractional distillation of crude petroleum.

Step 3: Fractional distillation

In an iron retort, the crude oil is heated to about 400-430°C. Here, all volatile matter are evaporated, components which are not volatile like tar and asphalt are settled at the bottom of the column. The hot vapours are then passed through a distillation column, shown in figure.

The distillation chamber is a steel cylindrical tube about 31 m height and 3 m in diameter and inside the chamber, trays are fitted at short distances. Every tray is having many holes and an up going short tube with a bubble cap. At different heights of chamber, the vapours go up, begin to cool and condense in fractions. Fractions which are having higher boiling point condenses first and lower boiling fractions condense one after the other.

2.3.1 Cracking, Synthetic petrol

Cracking

Cracking is a process by which the hydrocarbons of high molecular mass are decomposed into hydrocarbons of low molecular mass by heating in the presence or absence of a catalyst. Generally, aluminum silicates are used as catalyst.

Example:

$$\underset{\text{decane}}{C_{10}H_{22}} \xrightarrow{\text{cracking}} \underset{\text{(n-pentane + Pentene)}}{C_5H_{12} + C_5H_{10}}$$

Gas, oil and kerosene contain hydrocarbons of high molecular mass and boiling point. They are unsuitable as fuel in automobiles. Hence they are decomposed into hydrocarbons of low molecular mass and low boiling point.

Fluidized (Moving) Bed Catalytic Cracking

Principle

In this the finely divided catalyst is kept agitated by cracking oil, so that it can be handled like a fluid system. One of the advantages of FCC is that it gives very good contact between oil and catalyst. Therefore, high yield of petrol is obtained.

- Conditions: Feed stalk: Vapours of heavy oil fraction.
- Catalyst: $Al_2O_3 + SiO_2$ (Alumina + silicon dioxide)
- Temperature: 530°C
- Pressure: Little above the normal pressure.
- Yield: Very high, usually 10 gallons per day.

The finely divided catalyst is fluidized by the upward passage of the feed stalk in the cracking chamber. Cracked vapours are continuously withdrawn from the cracking chamber and fed into fractionating column where it gets separated into gas and gasoline. The uncracked oil may be cracked in the second stage of cracking process.

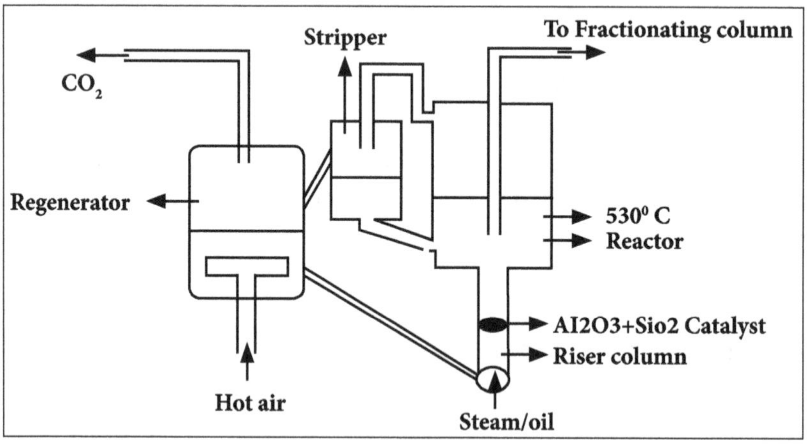

The spent catalyst is continuously withdrawn from the bottom of the cracking chamber and transported into regenerators by a stream of air in which the carbon deposited on the catalyst is burnt off using hot air. The regenerated catalyst is mixed with fresh feed stalk and returned to cracking chambers.

Synthetic Petrol

Manufacture of Synthetic Petrol (Bergius Process)

Coal is grounded and made into a paste with heavy recycle oil and a catalyst like tin oleate. The paste is sent along with H_2 at 250-350 atm pressure into a converter which is maintained at 450°C-500°C temperature. The un-reacted coal is filtered-off and the liquid product is distilled. Hydrogen combines with coal to form saturated hydrocarbons which decompose at high temperatures yielding low boiling hydrocarbons.

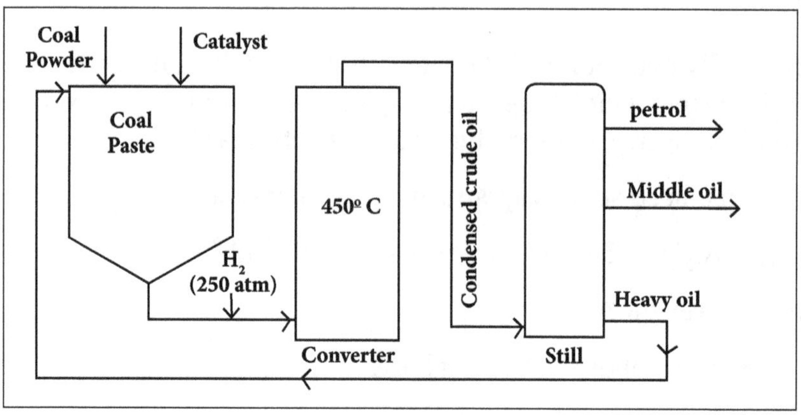

Bergius process.

The crude oil is fractionated to get, gasoline, middle oil and heavy oil which is recycled. Middle oil is hydrogenated in vapour phase with catalyst to yield more gasoline. Yield of gasoline is 60% of the coal dust.

Synthesis of Petrol by Fischer-Tropsch Process (Indirect Conversion of Coal)

This method involves the following steps.

Production of water gas

Water gas $(CO + H_2)$ is obtained by passing steam over white hot coal.

$$\underset{\text{Coke steam}}{C + H_2O} \xrightarrow{1200\,^{\circ}C} \underset{\text{Water gas}}{CO + H_2}$$

The water gas is mixed with hydrogen and the mixture is purified by passing through Fe_2O_3 and then passed into a mixture of $Fe_2O_3 + Na_2CO_3$. Fe_2O_3 is used to remove H_2S. $Fe_2O_3 + Na_2CO_3$ is used to remove organic sulphur compounds.

2.3.2 Petrol Knocking, Diesel knocking

Knocking

The explosive combustion of petrol and air mixture produces shock waves in I.C. engines which hit the walls of the cylinder and piston, producing a rattling sound is termed as knocking.

Mechanism of Knocking

Beyond a particular compression ratio the petrol mixture suddenly burns into flame. The rate of flame propagation raises from 20 to 25 m/s to 2500 m/s, which propagates very fast, producing a rattling sound. The activated peroxide molecules decomposes to give number of gaseous products which produces thermal shock waves.

The reactions of normal and explosive combustion of fuel can be given as follows taking ethane as an example.

$$C_2H_6 + 3\,\tfrac{1}{2}\,O_2 \rightarrow 2CO_2 + 3H_2O \text{ (Normal combustion reaction)}$$

$$C_2H_6 + O_2 \rightarrow CH_3-O-O-CH_3 \text{ (Explosive combustion reaction)}$$

$$CH_3-O-O-CH_3 \rightarrow CH_3-CHO + H_2O$$

$$CH_3-CHO + 1\,\tfrac{1}{2}\,O_2 \rightarrow HCHO + CO_2 + H_2O$$

$$HCHO + O_2 \rightarrow CO_2 + H_2O$$

Petrol Knocking

Knocking should be prevented in petrol engines. It occurs in four stroke engines when some hydrocarbons ignite too early. It causes a knocking sound and reduces the engine performance.

We measure the problem using the octane number. The higher the number, better the fuel. A fuel with a low octane number is likely to cause knocking. One with an octane number above 100 must be fine.

There are two main ways of increasing the octane number of petrol:

- Use aromatic alkanes with rings.
- Use alkanes with branching chains rather than straight chains.

These alkanes are produced on the site using isomerism and the reformer. They are then blended into petrol at the end. The blended petrol is constantly monitored for its octane number.

Diesel Knocking

Cetane number is used to measure knocking in diesel engine. Diesel knock is the rattling, clanking sound emitted from a running diesel engine. This noise is caused by the compression of air in the cylinders and the ignition of fuel as it is injected into the cylinder. This is the same as that of a gasoline engine suffering from spark knocking or pre-ignition. The timing of the fuel being injected into the diesel engine is critical to prevent parts breakage, which can result from severe knock.

A diesel engine functions differently than its gasoline counterpart. In a gasoline engine, the fuel is mixed with air and then compressed before an electric spark ignites the mixture. In a diesel engine, only the air is compressed. The fuel is then injected into the cylinder filled with compressed air and the heat from the compressed air ignites the fuel without the help of an electric ignition.

The telltale sound of an operating diesel engine is due to the fuel injection process. By injecting raw fuel into extremely hot compressed air, the fuel ignites as the piston is still travelling up in the cylinder, causing a detonation and subsequent rattling sound to be heard. The process is compression driven and higher the compression ratio within the cylinder, the greater the power output of the engine.

Gasoline engines operate at 8:1 to 10:1 compression ratios whereas, a diesel engine operates at 14:1 to 25:1 compression ratios. This higher compression ratio allows the diesel engine to operate much more efficiently than gasoline. Diesel knock is a byproduct of the raised compression and fuel injection process and is an acceptable result of the ignition sequence.

A diesel engine is difficult to start in cold weather due to lack of an electronic ignition system. Few manufacturers equip diesel engines with glow plugs to aid in starting the engine in cold weather. A glow plug uses the battery to heat a wire coil red hot in the combustion chambers. This causes more noticeable diesel knock in the engine till it reaches the operating temperature. Knocking declines as the fuel begins to ignite more easily within the engine.

Few manufacturers have created special engine mounts that helps muffle diesel knocks from passenger compartments. As the cost of fuel rises, diesel engines are being fitted in an increasing amount to passenger vehicles due to superior fuel efficiency.

2.3.3 Octane and Cetane ratings, and Anti-knock agents

Octane Number of a Fuel

Octane number of a fuel is a measure of its ability to resist knocking. The knocking characteristics of petrol are described by the octane number. Higher the octane number lowers the knocking tendency and better is the quality of petrol. The octane number is an arbitrary factor. Isooctane when used as a fuel found to have zero knocking hence its octane number is taken as 100. But when n-heptane was used as a fuel it had maximum knocking hence its octane number is taken as zero.

Octane number is equal to the percentage of volume of isooctane also known as 2,2,4-trimethyl pentane in a mixture of n-heptane and isooctane having the same knocking tendency compared to the sample of gasoline being tested. Isooctane has the best anti-knocking properties whereas, n-heptane has poor anti-knocking properties.

The hydrocarbons present influence the knocking properties of gasoline which vary according to the series: straight chain paraffin > branched chain paraffin > olefin > cycloparaffin > aromatics. The fuel which has same knocking tendency with the mixture having 80% isooctane has octane number as 80.

The most effective anti-knock agent added is tetraethyl lead along with ethylene dibromide which prevents the deposition of lead by forming volatile lead halides. Other anti-knocking agents are diethyl telluride, tertiary butyl acetate and tetra methyl lead.

1.0 - 1.5 ml of TEL is added per liter of petrol. TEL functions by being converted to a cloud of finely divided lead oxide particles, which reacts with any hydrocarbon peroxide molecules formed in the engine cylinder, thereby preventing knocking and solving the chain oxidation reaction.

Octane number is defined as the percentage of isooctane present in a mixture of isooctane and n-hectane.

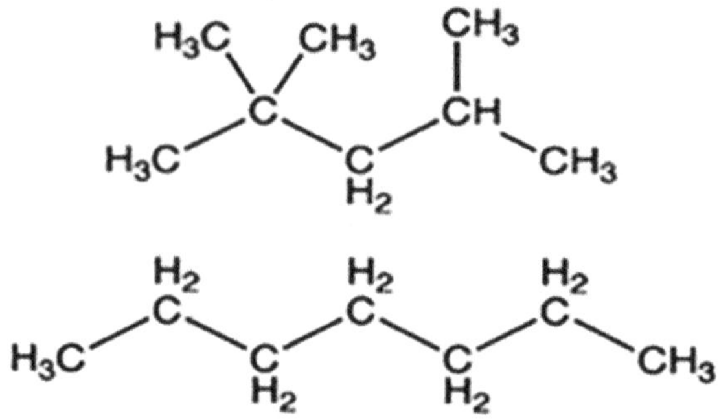

Octane number.

For a petrol sample under test, whose octane number is to be determined, is compared with reference standards of isooctane and n-heptane prepared at different ratios (90:10; 80:20, 75:25 etc.) and the knocking characteristics of these are determined under same conditions as that of the sample under test.

Suppose the knocking characteristics of the fuel are same as that of 80:20 mixtures, the octane number of the fuel is 80. Therefore, the octane number is defined as the percentage of volume of isooctane present in a standard mixture of isooctane and n-heptane which has the same knocking characteristics as the petrol under test.

$CH_3 - C(CH_3)_2 - CH_2 - CH(CH_3) - CH_3$	$CH_3 - (CH_2)_5 - CH_3$
Iso octane (2,2,4 Trimethyl Pentane)	**n-Heptane**
O N =100, Anti K value=100	**O N =0, Anti K value=0**

Cetane Number of a Fuel

Cetane number of a fuel is a measure of its ability to resist knocking. The knocking characteristics of diesel are described by the cetane number. Cetane number is defined as the percentage of volume of cetane or hexadecane, present in a standard mixture of cetane and α- methyl naphthalene which has the same knocking characteristics as the diesel under test.

Generally diesel fuels with cetane numbers of 70-80 are used.

$C_{16}H_{34}C_{10}H_7 - CH_3$

Cetane α- Methyl naphthalene

C N = 100, Anti K value = 100 C N = 0, Anti K value = 0

Prevention of Knocking (Anti-knocking Agents)

The substances added to gasoline to control knocking are called anti-knocking agents. Usually organometallic compounds are added to gasoline.

The followings are the substances to prevent knocking in I.C. engines:

- TEL - Tetra Ethyl led (leaded petrol).
- TML - Tetra Methyl led.
- MTBE - Methyl Tertiary Butyl Ether (unleaded petrol)

Leaded Petrol

The petrol containing TML or TEL as the anti-knocking agents is known as the leaded petrol. TML or TEL are very good anti-knocking agents but has some disadvantages as follows:

- It spoils the catalyst used in catalytic converter.
- The lead is a poisonous air pollutant.
- When the combustion lead is deposited as lead oxide on the piston and engine walls, it leads to mechanical damage.

Unleaded Petrol

The petrol which contains anti-knocking agent other than lead is known as unleaded petrol.

Example: MTBE is used as an anti-knocking agent in place of TEL or TML and the petrol is known as unleaded petrol.

2.4 Power Alcohol

Alcohol is an excellent alternative motor fuel for gasoline engines. Alcohols are fuels belonging to the oxygenates family having one or more oxygen which contributes to the combustion. Only two of the alcohols such as methanol and ethanol are economically and technically suitable as fuels for internal combustion engines.

If about 20-25% ethyl alcohol is blended with petrol and is used as a fuel then it is known as power alcohol for internal combustion engine. Gasohol is a mixture of 90% unleaded gasoline and 10% ethyl alcohol or ethanol. Its performance as a motor vehicle fuel is comparable to that of 100% unleaded gasoline, with the added benefit of superior anti-knock properties.

No engine modifications are needed for the use of gasohol, which has in recent years gained some acceptance as an alternative to pure gasoline. Methanol is produced by a variety of process, the most common are, distillation of wood, distillation of coal, natural gas and petroleum gas. Ethanol is produced mainly from biomass transformation or bioconversion.

Advantages of Power Alcohol

Alcohol has characteristics which makes it as a natural engine fuel. For instance:

- It has a high octane rating, which prevents engine detonation under load.
- It burns clean, in fact, not only the noxious emissions are drastically reduced, but also the internal parts of the engine are purged of gum deposits and carbon which, of course, do not build up as long as alcohol is used as a fuel.
- An alcohol burning engine tends to run cooler than its gasoline powered counterpart thus, extending engine's life and reducing the chance of overheating.
- It has less starting problems.
- Alcohol removes all traces of moisture in the petrol.

2.4.1 Biodiesel

Biodiesel is an oxygenated, renewable fuel made from a variety of agricultural resources such as soybeans or rapeseeds. Biodiesel is a biodegradable, non-toxic replacement for petrol and diesel.

Biodiesel is made from natural, renewable sources such as new and used vegetable oil and animal fats. Chemically biodiesel is described as a fatty acid mono alkyl ester.

Through a process called esterification, oils and fats are reacted with CH_3OH and NaOH catalyst at 500°C to produce fatty acid methyl ester along with the co-products such as glycerin, glycerin bottoms, soluble potash and soaps.

$$\text{Rape seed oil} + CH_3OH \rightarrow \text{Biodiesel} + \text{co-products}$$

Using biodiesel in a conventional diesel engine substantially reduces emissions of carbon monoxide, nitrated polycyclic aromatic hydrocarbons, polycyclic hydrocarbons, sulphates, unburnt hydrocarbons.

One of the primary advantages of biodiesel is its renewability. As a renewable domestic energy source, biodiesel can help reduce dependence on petroleum imports.

Biodiesel is non-toxic, biodegradable and suitable for sensitive environments. Biodiesel contains no petroleum but can be blended at any level with petrol and diesel to create a biodiesel blend. Biodiesel refers to the pure alternative fuel before blending with petroleum based diesel fuel.

Advantages

- Substitute for petroleum which is a non-renewable resource.
- Leads to reduction in CO_2 emission by 75%.
- Complete burning of bio-diesel leading to improved emission.
- Biodegradable.

2.4.2 Gaseous Fuels

Gaseous fuels occur in nature besides being manufactured from liquid and solid fuels. Gas fuels are the most convenient, because they require the least amount of handling and are used in the simplest and the most maintenance free burner systems.

Types of Gaseous Fuel

The following is a list of the types of gaseous fuels:

1. Fuels naturally found in nature:
   - Methane from coal mines.
   - Natural gas.
2. Fuel gases made from solid fuels:
   - Gases derived from waste and biomass.
   - Gases derived from coal.
   - From other industrial processes.
3. Gases made from petroleum:
   - Gases from oil gasification.
   - Liquefied Petroleum Gas.
   - Refinery gases.

Advantages

Gaseous fuels due to the ease and flexibility of their applications shall possess the following advantages over solid or liquid fuels.

- It can be conveyed easily through pipelines to the actual place of need, thereby eliminating manual labour in transportation.
- It can be lighted at ease.

- It has high heat contents and hence, helps us in having higher temperatures.
- It can be pre-heated by the heat of hot waste gases, thereby affecting economy in heat.
- Its combustion can readily by controlled for change in demand like oxidizing or reducing atmosphere, length flame, temperature, etc.
- It is clean.
- It does not require any special burner.
- It burns without any smoke and ashes.
- It is free from impurities found in solid and liquid fuels.

Disadvantages

- They are highly inflammable, so chances of fire hazards in their use is high.
- Very large storage tanks are needed.

2.4.3 Natural gas, LPG and CNG

Natural Gas

The natural gas is said to be a fossil fuel formed during layers of buried plants, gases and animals that are exposed to intense heat and pressure over thousands of years. Therefore, the energy that the plants originally obtain from the sun is stored in the form of chemical bonds in natural gas.

Characteristics of Natural Gas

Such as crude oil, natural gas is an energy source based on hydrocarbon chain but, the composition of natural gas is generally different than the composition of crude oil. The natural gas is primarily composed of methane, even though some natural gas deposits additionally contained substantial fractions of other hydrocarbon liquids or gases like ethane and propane. Most of the gas deposits also contains impurities like sulphur or other carbon compounds that should be separated prior to the gas being injected into the transmission or distribution pipelines. The gas deposits that consists primarily of methane are called as dry gas deposits, while those with larger fractions of other hydrocarbons are termed as wet or rich gas deposits.

Unlike oil, natural gas is essential to its transportation system without pipelines, there is no economical way to get large quantities of gas to market. Natural gas pipelines generally require dedicated assets. Using oil or petroleum product pipelines to transport natural gas is not really possible and transporting other products in natural gas pipelines is also not possible without completely re-purposing the pipeline.

Chemical Composition of Natural Gas

The natural gas is primarily composed of methane, ethane, propane and also heavier hydrocarbons. It contains small amounts of nitrogen, carbon dioxide, hydrogen sulphide and also traces of water.

Applications

- The gas is used in cooling and co-generation, which is useful in producing electricity. Then, it is used as chilled water for the cooling system in an office building. The application of gas cooling can be seen in overseas. For instance, the new Tokyo national airport in Japan, Kuala Lumpur international airport and Suvarnabhumi airport, Thailand.
- The natural gas can be used for cooking by replacing LPG in hotels, hospitals and residences. The gas will be used with all types of stoves, ovens, grills and rice cookers.
- The natural gas is used for producing hot water in hotels, the laundry services, the sterile process at hospitals and household residence.

LPG and CNG

CNG

It is stored in a high pressure container usually at 3000 to 3600 psi. CNG is mainly used as an alternative fuel for internal combustion engines such as automobiles engines. This generates low hydrocarbon emissions and a significant quantity of nitrogen oxide emissions.

LPG

The Liquefied Petroleum Gas or Liquid Petroleum Gas also referred to as propane or butane, are the flammable mixtures of hydrocarbons, which is used as fuel in heating appliances, cooking equipments and vehicles.

It is used as an aerosol propellant and also as a refrigerant replacing chlorofluorocarbons in an effort to reduce damage to the ozone layer. It is specifically used as a vehicle fuel and often referred to as autogas.

2.5 Combustion

Combustion is a method where the rapid chemical combination of oxygen with the combustible elements of fuel, results in the production of heat and light. Combustion is accomplished by mixing the fuel and air at elevated temperatures. Complete combustion of fuel is possible in the presence of adequate supply of oxygen. The air supplies oxygen which unites chemically with carbon, hydrogen and a few minor elements in the fuel to produce heat. Steam has been generated from the burning of a variety of fuels.

The heating value of a fuel may be determined either by burning a sample in the calorimeter or by calculating from chemical analysis. The calculation must be based on ultimate analysis that reduces the fuel to its elementary constituents of ash, carbon, hydrogen, moisture, nitrogen, oxygen and sulphur to secure a reasonable degree of accuracy. The proximate analysis which determines only the percentage of moisture, volatile matter, fixed carbon and ash without determining the ultimate composition of the volatile matter cannot be used for computing the heat of combustion with the same degree of accuracy as an ultimate analysis but estimates may be based on the ultimate analysis.

Oxygen (O_2) is one of the most common elements on the earth making up to 20.9% of our air. Rapid fuel oxidation results in large amount of heat. Solid or liquid fuels must be changed to gas before they burn. Generally, heat is required to change liquids or solids into gases. Fuel gases will burn in their normal state if enough air is present.

Most of the 79% of air is nitrogen, with traces of other elements. Nitrogen is considered to be a temperature reducing diluent that must be present to obtain the oxygen required for combustion.

Nitrogen reduces the combustion efficiency by absorbing heat from the combustion of fuels and diluting the flue gases. This reduces the heat available for transfer through the heat exchange surfaces. It also increases the volume of combustion byproducts, which have to travel through the heat exchanger and up the stack, faster to allow the introduction of the additional air-fuel mixture. This nitrogen can combine with oxygen to produce oxides of nitrogen (NO_x), which are known as toxic pollutants.

Carbon, hydrogen and sulphur in the fuel combines with oxygen in the air to form carbon dioxide, water vapour and sulphur dioxide, releasing 8084 kcals, 28,922 kcals and 2224 kcals of heat respectively.

Under certain conditions, carbon may also combine with oxygen to form carbon monoxide, which results in the release of a smaller quantity of heat. Carbon burned to CO_2 will produce more heat per pound of fuel than when CO or smoke is produced.

$$C + O_2 \rightarrow CO_2 + 8084 \text{ kcals/kg of carbon}$$

$$2C + O_2 \rightarrow 2\,CO + 2430 \text{ kcals/kg of carbon}$$

$$2H_2 + O_2 \rightarrow 2H_2O + 28{,}922 \text{ kcals/kg of hydrogen}$$

$$S + O_2 \rightarrow SO_2 + 2224 \text{ kcals/kg of sulphur}$$

Each kilogram of CO formed means a loss of 5654 kcal of heat.

2.5.1 Calculation of Air for the Combustion of a Fuel

Theoretical Oxygen

It is the amount of oxygen needed for complete combustion of all the fuel fed to the combustion chamber, assuming that all carbon in the fuel reacts with oxygen, to form CO_2, all hydrogen reacts to form H_2O and all sulphur reacts to form SO_2.

Theoretical Air

It is the quantity of air that contains theoretical oxygen, (i.e.) it is the minimum quantity of air required to burn the fuel completely so that all carbon gets converted into CO_2, all hydrogen into H_2O and all sulphur into SO_2.

Theoretical air requirement of any fuel is calculated as,

$$\text{Theoretical air requirement in moles} = \frac{\text{Theoretical oxygen requirement in moles}}{0.21} \quad \ldots(1)$$

In actual combustion practice, air is used in excess of that theoretically required to assure complete combustion.

Excess Air

It is the amount by which the air fed to the combustion chamber exceeds the theoretical air.

$$\text{Percent excess air} = \left(\frac{\text{actual air supply} - \text{theoretical air requirement}}{\text{theoretical air requirement}}\right) \times 100 \quad \ldots(2)$$

$$\%\ \text{excess oxygen} = \%\ \text{excess air}$$

The quantities of actual air supplied and theoretical air required can be used either in weight units or in molar units in equation (2).

The actual air supply can be obtained by knowing the % excess and theoretical air requirement by the relation,

$$\text{Actual air supply} = \text{Theoretical air requirement}\left[1 + \frac{\%\ \text{excess}}{100}\right]$$

If we know the feed rate of a fuel and the stoichiometric equation for complete combustion of the fuel, we can calculate the theoretical oxygen demand and air feed rate. The equation for the calculation of % excess air is applicable for the calculation of % excess oxygen. The % excess air and % excess oxygen are one and the same terms.

The theoretical air required to burn a given quantity of fuel does not depend on how much fuel is actually burned and the value of the percent excess air depends only on the theoretical air and the air supply rate and not on the quantity of O2 actually consumed or whether the combustion is complete or partial. The theoretical as well as actual air requirements are expressed in different units, example, kg per kg fuel or m3 per kg fuel. For converting the amount of air expressed in kmol into the amount of air in kg, the average molecular weight of air can be taken as 29. For converting the amount of air from kmol/mol into l/m3, we have to use the ideal gas law.

The excess air requirement depends upon the type of fuel burnt. Gaseous fuels require very less excess air. Liquid fuels require more excess air than gaseous fuel and solid fuels require higher amounts of excess air than liquid fuels. Gaseous fuels are burnt with 5 to 15% of excess air while liquid and solid fuels are burnt with 10 to 50% of excess air.

Problems

1. A fuel contains 80% CH_4 and 20% C_2H_6 by volume. Let us determine the number of atoms in kmol and let us also show that the number of atoms on the L.H.S. balances the number of atoms on the R.H.S.

Solution:

Given:

Fuel contains 80% CH_4 and 20% C_2H_6

We could choose any quantity of fuel but for convenience we shall use 1 k mole, which can be written as,

$$\left(0.8\,CH_4 + 0.2\,C_2H_6\right)$$

This mixes with Y k moles of air and then the combustion process goes to completion.

Now the composition of air is 79% N_2 and 21% O_2.

Therefore, 1 k mole of air may be written as,

$$\left(0.79\,N_2 + 0.21\,O_2\right)$$

We may therefore write the combustion equation as,

$$\left(0.8\,CH_4 + 0.2\,C_2H_6\right) + Y\left(0.79\,N_2 + 0.21\,O_2\right) \rightarrow a\,CO_2 + b\,H_2O + c\,O_2 + d\,N_2$$

(The reactants) (The products of complete combustion)

Because the combustion process does not create or destroy atoms, the number of atoms on the L.H.S. of the equation should exactly balance the number of atoms on the R.H.S.

Number of atoms in k moles		
L.H.S	R.H.S	
Carbon C	0.8 + 2 x 0.2	a ....(i)
Oxygen O	2 x 0.21 x Y	2a + b + 2c ....(ii)
Hydrogen H	4 x 0.8 + 6 x 0.2	2b ....(iii)
Nitrogen N	2 x 0.79 x Y	2d ....(iv)

Solving equations (i) to (iv) simultaneously we get,

From (i), a = 1.2

From (iii), b = 2.2

From (ii), c = 0.21 Y - 2.3

From (iv), d = 0.79 Y

2.5.2 Flue Gas Analysis and Orsat Apparatus

The mixture of gases like CO_2, CO, O_2 etc., coming out from the combustion chamber is called as flue gas. The amount of flue gases like CO_2, CO, O_2 etc., can be estimated by using Orsat's apparatus on the absorption principle. The above gases are absorbed by KOH, alkaline pyrogallol and ammoniacal cuprous chloride solutions respectively.

Analysis of these gases gives an idea about the complete combustion of fuel.

- If CO is high in the flue gas then, it shows incomplete combustion of fuel and short supply of oxygen.
- If CO_2 and O_2 are high in the flue gas then, it shows complete combustion of fuel and excess supply of oxygen.

It consists of a horizontal tube having three way stop cock at one end. The end of a three way stop cock is connected to a U tube containing fused $CaCl_2$ to remove moisture in the gas. The end of the tube is connected with a graduated burette. The burette is surrounded by a water jacket in order to keep the temperature of gas constant. The lower end of the burette is connected by a water reservoir by means of rubber tube. The level of the water in the burette can be raised or lowered by raising or lowering the reservoir. The middle of the horizontal tube is connected with three bulbs (1, 2 and 3) for absorbing the flue gases as follows:

- Bulb '1' contains KOH solution and it absorbs only CO_2.
- Bulb '2' contains alkaline pyrogallol solution and it absorbs only O_2.

- Bulb '3' contains ammoniacal cuprous chloride solution and it absorbs CO.

Working of Orsat Apparatus

Now the three way stop cock is opened and the burette is filled with water by raising the water reservoir to remove air from the burette. Then, the flue gas is taken in the burette up to 100 cc by raising and lowering the reservoir. The three way stop cock is now closed.

Absorption of Gases in Bulbs

1. Absorption of CO_2

The stopper of bulb '1' is opened and the flue gas is allowed to pass by raising the water reservoir. CO_2 present in the flue gas is absorbed by KOH. This process is repeated several times by raising and lowering the water reservoir until the volume of burette becomes constant. The decrease in volume of burette indicates the volume of CO_2 in 100 cc of the flue gas. Now the stopper of the bulb '1' is closed.

2. Absorption of O_2

The stopper of the bulb '2' is opened and the flue gas is allowed to pass. O_2 present in the flue gas is absorbed by alkaline pyrogallol. This process is repeated several times until the volume of burette becomes constant. The decrease in volume of flue gas in burette indicates the volume of O_2 . Now the stopper of bulb '2' is closed.

3. Absorption of CO

The stopper of the bulb '3' is opened and the flue gas is allowed to pass. 'CO' present in the flue gas is absorbed by ammoniacal cuprous chloride solution. This process is repeated several times until the volume of burette becomes constant. The decrease in volume of flue gas in burette indicates the volume of CO. The remaining gas in the burette after the absorption of CO_2, O_2 and CO is taken as nitrogen.

$$\text{The \% of } N_2 = \left[100 - \left(\text{\% of } CO_2 + \text{\% of } O_2 + \text{\% of CO}\right)\right]$$

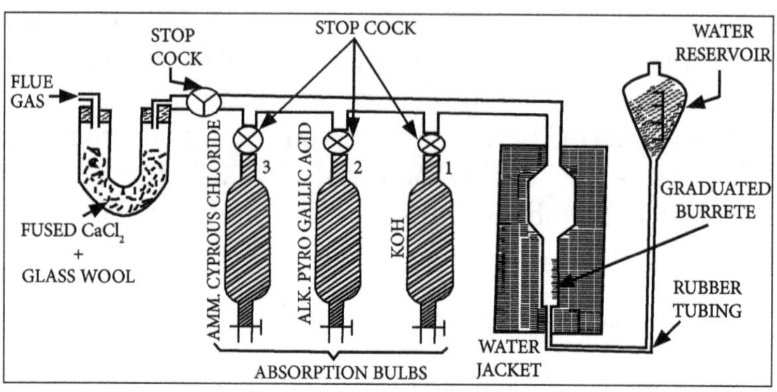

Flue gas analysis.

2.5.3 Numerical Problems on Combustion

1. Let us calculate the volume of air required for complete combustion of 1 m^3 of gaseous fuel having the composition: CO = 46%, CH_4 = 10%, H_2 = 4%, C_2H_4 = 2%, N_2 = 1% and the remaining being CO_2.

Solution:

Given:

CO = 46%, CH_4 = 10%, H_2 = 4%, C_2H_4 = 2%, N_2 = 1%

1. 1 m^3 of fuel contains,

4/100 = 0.04 m^3 of H_2.

10/100 = 0.10 m^3 of CH_4.

1/100 = 0.01 m^3 of N_2.

46/100 = 0.46 m^3 of CO.

2/100 = 0.02 m^3 of C_2H_4.

2. N_2 and CO_2 are non-combustible constituents hence, they do not undergo combustion.

3. The combustion equation can be written as follows:

- $H_2 + \frac{1}{2} O_2 \rightarrow H_2O$
    - 1 m^3 of H_2 requires 0.5 m^3 of O_2.
    - 0.04 m^3 of H_2 requires = 0.5 × 0.04/1 = 0.02 m^3 of O_2.
- $CH_4 + 2O_2 \rightarrow CO_2 + 2H_2O$
    - 1 m^3 of CH_4 requires 2 of O_2.
    - 0. 10 m^3 of CH_4 requires = 0.10 × 2/1 = 0.2 m^3 of O_2.
- $CO + \frac{1}{2} O_2 \rightarrow CO_2$
    - 1 m^3 of CO requires 0.5 m^3 of O_2.
    - 0.46 m^3 of CO requires = 0.46 × 0.5/1 = 0.23 m^3 of O_2.
- $C_2H_4 + 3O_2 \rightarrow 2CO_2 + 2H_2O$
    - 1 m^3 of C_2H_4 requires 3 m^3 of O_2.
    - 0.02 m^3 of C_2H_4 requires = 0.02 × 3/1 = 0.06 m^3 of O_2.

- Total volume of O_2 required = 0.02 + 0.2 + 0.23 + 0.06 = 0.51 m^3 of O_2.

We know that, 21 m^3 of O_2 is supplied by 100 m^3 of air.

- 0.51 m^3 of O_2 is supplied by = 100 × 0.51/21 = 2.428 m^3 of air.
- Amount of oxygen required by 100 m^3 of fuel = 2.428 m^3 of air.

2. A fuel contains C = 75%, H = 4%, O = 5%, S = 7% remaining ash. Let us calculate the minimum quantity of air required for complete combustion of 1kg of fuel.

Solution:

Given:

Weight of the fuel = 1 kg

Weight of C in the fuel = 0.75 kg

Weight of H in the fuel = 0.04 kg

Weight of O in the fuel = 0.05 kg

Weight of S in the fuel = 0.07 kg

- $CO + O_2 \rightarrow CO_2$
  - 12 kg of carbon requires 32 kg of oxygen.
  - 0.75 kg carbon requires = (32/12) x 0.75 = 2 kg.
- $H_2 + \frac{1}{2} O_2 \rightarrow H_2O$
  - 2 kg of hydrogen requires 16 kg of oxygen.
  - 0.04 kg hydrogen requires = (16/2) x 0.04 = 0.32 kg.
- $S + O_2 \rightarrow SO_2$
  - 32 kg of carbon requires 32 kg of oxygen.
  - 0.07 kg carbon requires = (32/32) x 0.07 = 0.07 kg.
  - Total weight of oxygen required = 2 + 0.32 + 0.07 = 2.39 kg.

But,

- Weight of oxygen already present = 0.05 kg.
- Actual weight of oxygen required = 2.39 - 0.05 = 2.34 kg.
- Weight of air required = 2.34 x (100/23) = 10.17 kg of air.

3. Let us calculate the mass of air required for complete combustion of 5 kg of coal, containing H = 15%, C = 80%, O = rest.

Solution:

Given:

5 kg of coal contains, C = 4 kg, H = 0.75 kg, O = (5 - 4 - 0.75) kg = 0.25 kg.

Amount of air required for complete combustion of 5 kg coal = [5 x (32/12) + 0.75 x (16/2) - 0.25] kg x (100/23)

= [13.333 + 6.000 - 0.25] kg x (100/23) = 82.97 kg

4. A sample of coal was found to contain, C = 80%, H = 5%, O = 1%, N = 2% remaining being ash. Let us calculate the amount of minimum air required for complete combustion of 1 kg of coal sample.

Solution:

Given: C = 80%, H = 5%, O = 1%, N = 2%

Combustion reaction	Weight of air required
$C + O_2 \rightarrow CO_2$ 12 32	800 g (32/12) = 2146 gm 50 g (16/2) = 400 gm
$2H + 0.5O_2 \rightarrow H_2O$ 2 16	Total = 2546 gm Less O_2 in fuel = 10 gm Net O_2 required = 2536 gm

Weight of air required = 2536 g(100/23) = 11026 gm = 11.026kg.

5. Let us calculate the weight and volume of air required for the combustion of one kg of carbon.

Solution:

Carbon undergoes combustion according to the equation.

$$C + O_2 \rightarrow CO_2$$

12 32

Thus wt. of O_2 required for combustion of 12 gm of C = 32 gm.

Hence weight of oxygen required by 1 kg of carbon $= \frac{32}{12} \times 1 = 2.667$ kg

$$\therefore \text{ wt. of air (containing 23\% oxygen) required } = \frac{100}{23} \times 2.667 = 11.59 \text{ kg}$$

Now since 32 gm of oxygen occupied 22.4 litres at NTP

$$\therefore 2.667 \times 1000 \text{ gm of } O_2 \text{ will occupy } = \frac{22.4}{32} \times 2.667 \times 1000 = 1866.9\text{L}$$

So, volume of air (containing 21% oxygen) required

$$= \frac{100}{21} \times 1866.9 = 8890 \text{ Litres} = 8.89 \text{ m}^3$$

2.6 Explosives: Introduction and Classification

When a substance or a mixture is subjected to thermal or mechanical shock, it gets rapidly oxidized exothermically into the products of greatly increased volume, with a sudden release of potential energy called explosive.

Power to weight ratio of an explosive is nothing but the amount of power available from a given weight of explosive.

Characteristics of Explosives

- It should be cheap and stable under normal conditions.
- It must have at least one chemical bond that can be easily broken or, the molecule should have a low energy of dissociation.
- The rate of decomposition should be fast to produce a large volume of gaseous products exothermically.
- It should have a positive oxygen balance. The oxygen balance is expressed as a percentage surplus or deficiency of oxygen by weight. The oxygen balance indicates the oxygen contained in the molecule, which can be utilized to oxidize the C and H to CO_2 and H_2O respectively.

Classification of Explosives

Explosives are usually classified into three broad gro ups. They are as follows:

- Primary explosives (detonators).
- Low explosives (propellants).
- High explosives.

Uses of Explosives

Explosives can he used for constructive as well as destructive purposes. They can be used:

- For blasting of ores of iron and other metals.
- For breaking down coal, quarrying granite, limestone etc.
- In formation of tunnels, roads etc., which require blasting.
- As an ammunition in war.
- To manufacture bombs, grenades etc.
- For launching satellites with the help of rockets.

Applications of Explosives

Objects at a smaller distance can be shattered with high brisance explosives, explosives with low brisance can be used as propellants. (i.e.) shells, bombs and mines need high brisance explosives while all propellants belong to low explosives category.

For sub-soil blasting, the explosives are burst a few feet below the land. In this process, the sub-soil is brought up to the surface and the liberated gases such as SO_2, CO, etc., can kill the insects. If the explosives are nitrate-based, some fertilizers also get incorporated into the soil.

An explosive attached to a piece of metal is used to shape the metal to the required shapes that are not achievable by other methods. This kind of metal forming by explosives is of recent origin.

The explosives for dislodging the coal seam in coal mines must be carefully chosen. A long lasting flame may cause dangerous explosions due to exothermic burning of methane and coal dust. The addition of NaCI and AN to GTN not only helps to lower the flame temperature but also the duration of the flame. A mixture known as permitted explosive is used in coal mines.

2.6.1 RDX, TNT and Ammonium Nitrite - Rocket Fuels

RDX

RDX $(O_2NNCH_2)_3$ is an organic compound. It is a white solid generally used as an explosive. Chemically, it is classified as nitramide. A more powerful explosive than TNT, was widely used in World War II. RDX is also known as Research Department Formula X.

It is generally used in mixtures with other explosives and phlegmatizers or plasticizers. RDX is stable in storage and is considered as one of the most powerful and brisance material.

TNT

Trinitrotoluene $C_6H_2(NO_2)_3CH_3$ is also known as 2, 4, 6-trinitrotoluene. TNT is used to generate charge transfer salts. It is a yellow coloured solid, generally used as a reagent in chemical synthesis, but it is best known as an explosive material with convenient handling properties. The explosive yield of TNT is considered to be the standard measure of bombs and other explosives.

TNT is not the same as dynamite, a special formulation of nitroglycerin for use as an industrial explosive. It is prepared by nitration of toluene by nitrating agent in a reaction tank.

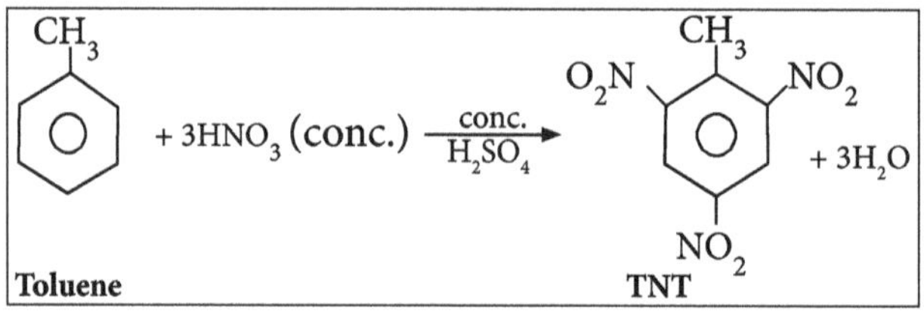

When TNT crystallizes out, the products are washed with ammoniacal solution of Na_2SO_3 and then with cold water. Now, the crystals of TNT are filtered and purified by melting.

Ammonium Nitrate

Ammonium nitrate is the nitrate salt of ammonium cation. It has the chemical formula NH_4NO_3, which is simplified to $N_2H_4O_3$. It is a white crystalline solid and is highly soluble in water. It is used in agriculture as a high nitrogen fertilizer. It is also used as a component in explosive mixtures, which is used in civil construction, mining and quarrying. It is a the major constituent of ANFO, a popular industrial explosive in North America, similar formulations have been used in improvised explosive devices.

Explosive Range

Explosive range is defined as the concentration of a substance in air between the Lower Explosive Limit and the Upper Explosive Limit. In this range, the substance will readily ignite.

Lower Explosive Limit

LEL is the minimum concentration of a substance in air which is required for ignition. Concentrations below the LEL will not ignite. Below LEL, the mixture is known as lean. If the concentration of a flammable gas or vapour is greater than 10% of the LEL, evacuate the area.

Upper Explosive Limit

UEL is the maximum concentration of a substance in air which is required for ignition.

Concentration above the UEL will not ignite. Above the UEL, the mixture is termed as rich.

Rocket Fuels

Modern rocket Fuel come in two main categories such as:

- Liquid fuels.
- Solid fuels.

Liquid Rocket Fuels

Liquid fuel rockets commonly use liquid oxygen and either liquid hydrogen or kerosene. These combinations work well in space and down closer to the ground, resulting in a multitude of uses from the first stages of the Saturn V and Falcon rockets to the Space Shuttle's main engines used to get the orbit in position in space.

Advantages

- Liquid fuel boosters are more easily reusable.
- Variable thrust- the amount of fuel and the rate of burn can be changed in flight.

Disadvantages

- Fragile, many complex parts.
- Oxidizer must be kept extremely cold.

Solid Rocket Fuel

Solid rocket fuel is easier and cheaper to handle as we do not have to cool materials to cryogenic temperatures. This means that solid fuels are prevalent in military applications and also in the first stages of space rockets. Made from powdered aluminium and an oxidizer, this fuel is generally used in booster rockets to give the aircraft, that extra kick needed to lift off and make their way into space.

Advantages

- Very stable and durable.
- More thrust for a similar size rocket.

Disadvantages

- Cannot be turned off, once it starts burning it goes up until the fuel is used up.
- Fuel decomposes, must be replaced.

3

Electrochemical Cells and Corrosion

3.1 Galvanic Cells, Reversible and Irreversible Cells, Single Electrode Potential

Galvanic Cells

Galvanic cells are electrochemical cells in which the electrons are transferred due to redox reaction is converted to electrical energy.

The flashlight battery is an electrochemical cell. An electrochemical cell is something that changes the chemical energy to electrical energy. It contains two compartments each with an electrode submerged in an electrolyte.

The electrode is just a conductor usually a metal that connects to a non-metallic part of a circuit. An electrolyte is a fluid that conducts the electricity. One of the plates is positive and one of them is negative. Those two plates are called the cathode and the anode and they are connected by a wire that completes the circuit.

Often one metal strip in the battery is copper and it acts as a cathode which is the electrode where, the reduction takes place and electrons are gained. The other electrode is the anode often zinc where, the oxidation takes place and electrons are lost. Electrons leave the system from the anode and go into the wire.

Electrons move from the anode to the cathode through the wire.

So what happens is the electrons on the anode are lost and travels through the wire to the cathode where the electrons are gained. The solution of the two electrodes helps to conduct the electrons from the surface of the electrode to the wire.

The two electrodes are in separate compartments separated by a porous barrier or salt bridge. The barrier or bridge allows the ions in both solutions to move from one side to the other to prevent charge from building up on the electrodes.

The chemical reaction that occurs in the electrochemical cell is a reduction-oxidation reaction or redox reaction. As its name implies that reaction's made up of two parts: a reduction reaction where the electrons are gained and an oxidation reaction where electrons are lost.

Galvanic cells are electrochemical cells in which the electrons are transferred due to redox reaction and are converted into electrical energy. A galvanic cell consists of two half-cells with each half-cell contains an electrode.

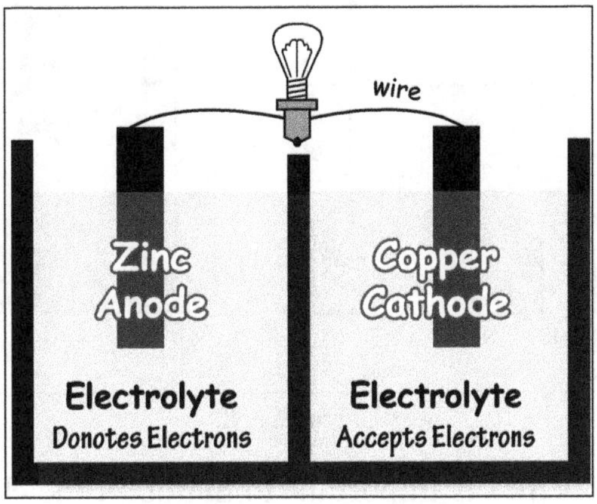

Electrochemical cell.

The electrode at which oxidation takes place is referred anode and the electrode at which reduction occurs is referred cathode. The electrons liberated to the electrolyte from the metal leaves the metal ions at anode.

The electrons from the solution are accepted by the cathode metal ion to become metal. Galvanic cell is generally represented as follows,

$$M_1 / M_1^+ || M_2^+ / M_2 \text{ or } M_1 / (\text{Saltof } M_1) || M_2 / (\text{Salt of } M_2)$$

Where, M_1 & M_2 are anode and cathode respectively and M_1^+ & M_2^+ are the metal ions in the respective electrolyte. The symbol || denotes salt bridge.

The above representation of galvanic cell is known as galvanic cell diagram.

Example: The typical example for galvanic cell is Daniel cell.

Setting Up Galvanic Cells

Galvanic cells harness the electrical energy available from the electron transfer during a redox reaction to perform useful electrical work. The key to gather the electron flow is to separate the oxidation and reduction half-reactions, connecting them by a wire, so that the electrons should flow through that wire. That electron flow is known as a current, may be sent through a circuit which could be part of any number of electrical devices like radios, watches, televisions, etc.

The figure below shows two typical setups for galvanic cells. The left hand cell diagram shows an oxidation and a reduction half-cell reaction joined by both a wire and a porous disk, while the right hand cell diagram shows the same cell substituting a salt bridge for the porous disk.

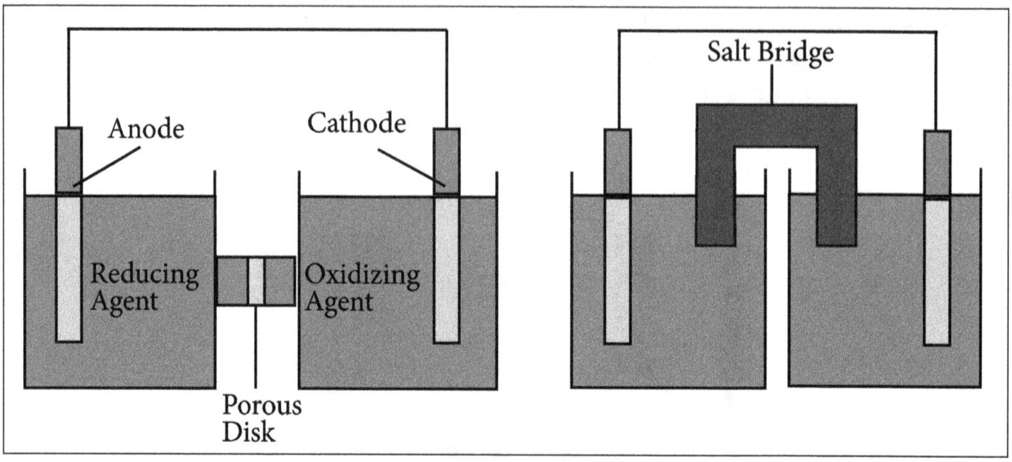

Galvanic cell.

The salt bridge or porous disk is important to maintain the charge neutrality of each of the half-cell by permitting the flow of ions with minimal mixing of the half-cell solutions. As the electrons are transferred from the oxidation half-cell to the reduction half-cell, a negative charge builds in the reduction half-cell and a positive charge in the oxidation half-cell. That charge buildup would serve to oppose the current from anode to cathode thus, effectively stopping the electron flow, if the cell lacked a path for ions to flow between the two solutions.

The above figure points out that the electrode in the oxidation half-cell is known as the anode and the electrode in the reduction half-cell is known as the cathode. A good mnemonic helps to remember is that "The Red Cat ate An Ox" meaning reduction takes place at the cathode and oxidation takes place at the anode.

The anode, as it is the source of the negatively charged electrons it is typically marked with a minus sign (-) and the cathode is marked with a plus sign (+).

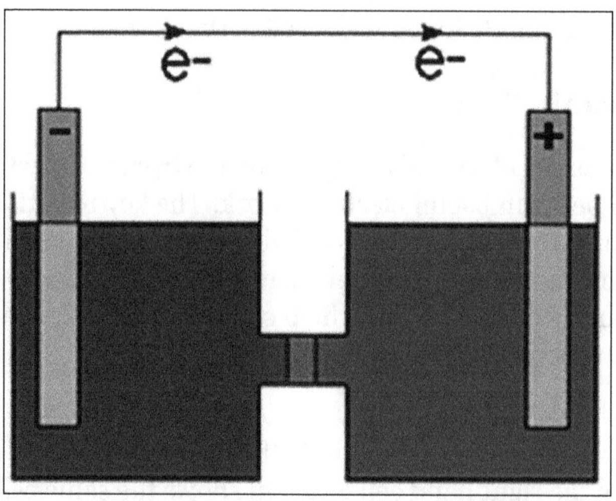

Galvanic cell showing direction of electron flow.

Physicists define the direction of current flow as the flow of positive charge based on an 18th century understanding of electricity. As we currently know, the negatively charged electrons flow in a wire. Therefore, the chemists indicate only the direction of electron flow on cell diagrams and not the direction of current. To make that point clear, the direction of electron flow is indicated with a arrow and the symbol for an electron is e^-.

Electrolytic Cell

Voltaic cells use a spontaneous chemical reaction to drive an electric current through an external circuit. These cells are important because they are the basis for the batteries that fuels to the modern society. But they aren't the only kind of electrochemical cell. It is also possible to construct a cell that does work on a chemical system by driving an electric current through the system. These cells are called electrolytic cells. Electrolysis is used to drive an oxidation-reduction reaction in a direction in which it does not occur spontaneously.

Electrolytic cells, like galvanic cells, are composed of two half-cells one is a reduction half-cell, the other is an oxidation half-cell. Though the direction of electron flow in an electrolytic cells may be reversed from the direction of spontaneous electron flow in the galvanic cells, the definition of both the cathode and anode remain the same reduction takes place at the cathode and oxidation occurs at the anode.

When comparing a galvanic cell to its electrolytic counterpart, as is done in, occurs on the right-hand half-cell. Because the directions of both the half-reactions have been reversed, the sign, but not the magnitude of the cell potential has been reversed. Note that copper is spontaneously plated onto the copper cathode in the galvanic cell whereas it requires a voltage greater than 0.78 V from the battery to plate the iron on its cathode in an electrolytic cell.

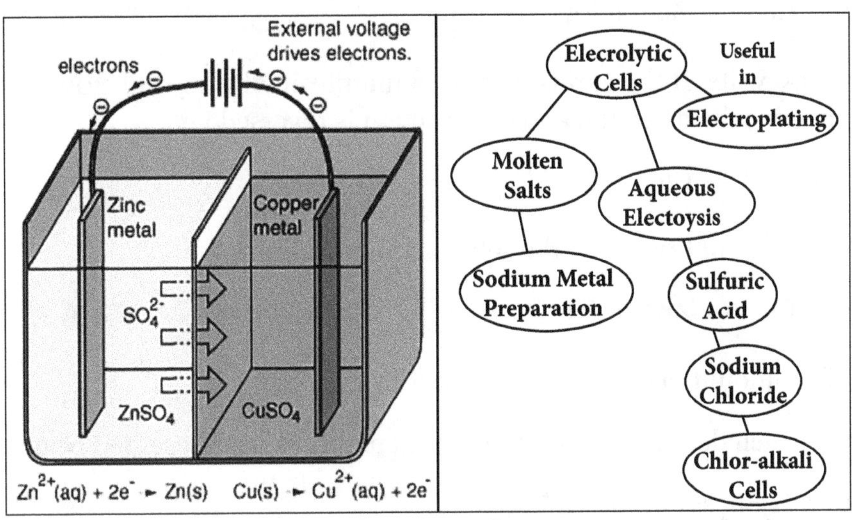

Electrolytic cell.

If a voltage greater than 1.10 volts is applied as illustrated to a cell under standard conditions, then the reaction,

$$Cu(s)+Zn^{2+}+(aq)\rightarrow Zn(s)+ Cu^{2+}+(aq)$$

Will be driven by removing Cu from the copper electrode and plating the zinc on the zinc electrode.

Electrolytic processes are very important for the preparation of pure substances like aluminum and chlorine.

Reversible and Irreversible Cells

- If the emf (exactly equal to its own) is applied by some external source, the cell reaction stops.
- If the external emf applied to it is slightly greater to its own end, the current begins to flow in opposite direction and the cell reaction is reversed.
- If the external emf is slightly less than that of the cell itself. A very small amount of the current will flow and a corresponding small amount of chemical reaction takes place in the cell.

When these requirements are satisfied by a cell, it is a reversible cell and if these conditions are not satisfied the cell is said to be irreversible. The difference between reversible and irreversible cells can be illustrated by taking two examples. First consider the Daniel cell,

$$Zn / ZnSo_4 // CuSo_4 / Cu$$

The net cell reaction is,

$$Zn + Cu^{2+} \rightarrow Zn^{2+} + Cu$$

The emf is 1.09 volts. If the emf is increased infinitesimally beyond 1.09 volts the current flows in opposite direction and cell reaction is reversed i.e.,

$$Cu + Zn^{3+} \rightarrow Cu^{2+} + Zn$$

If the emf is less than I 09 volts, the cell reaction continues,

$$Zn + Cu^{2+} \rightarrow Zn^{2+} + Cu$$

And a smaller amount of current is given out by the cell.

Another cell which does not satisfy the above condition is represented below,

$$Zn^{+} / H_2So_4 \text{ solution} / Ag^{+}$$

When the two electrodes are connected zinc dissolves with evolution of hydrogen to form zinc sulphate i.e.,

$$Zn(s) + H_2So_4 \rightarrow ZnSo_4 + H_2(g)$$

However, when the cell is connected with an external source of potential slightly greater than its own, silver dis-solves at one electrode, hydrogen is evolved at the other and the cell reaction becomes,

$$2Ag(S) + H_2So_4 \rightarrow Ag_2So_4 + H_2(g)$$

The cell however fulfills the first condition of revert ability but does not satisfy the second condition and hence it is an irreversible.

Single Electrode Potential

It is not possible to measure the absolute potential of a single electrode because neither the oxidation nor the reduction reaction can occur by itself. Also, we need two electrodes to measure the potential difference between two points. This difficulty can be solved by selecting one of the electrodes as a reference electrode and arbitrarily fixing the potential of that electrode as zero. Thus, it is possible to obtain the potential of an electrode, if the given electrode is coupled with another electrode having a potential of zero volts. The electrode whose potential can be taken as zero volts is known as primary reference electrode.

A standard hydrogen electrode has been assigned an electrode potential of 0.00 volt at all temperatures and is used as a primary reference electrode. The electrode potential of any other electrode is obtained by coupling it with a standard hydrogen electrode. The potential of the given electrode may be negative or positive with respect to the standard hydrogen electrode potential. The electrode potential represents the electron-accepting power of an electrode relative to that of the standard hydrogen electrode, i.e., the electrode potentials are reduction potentials.

The electrode potential depends upon:

- The nature of the metal and its ions.
- Concentration of the ions in the solution.
- Temperature.

Since the potential depends upon the concentration of the ions in solution we adopt the standard as 1 mole of ions per liter of solution. Electrode potentials under these standard conditions are called standard electrode potentials and are denoted as E°.

The standard conditions are:

- The concentration of the ions that take part in the electrode reaction should be 1 mol L^{-1}.
- For all the gaseous substances, which take part in the electrode reaction, the pressure should be equal to 1 atm.
- A temperature of 298 K.

Nernst Equation

The Nernst Equation is used to determine the single electrode potential of the cell. The Nernst Equation is derived from the emf and the Gibbs energy under non-standard conditions. Under standard conditions, the Gibbs free energy equation is,

$$\Delta G^{o} = -nFE^{\circ}.$$

Since,

$$\Delta G = AG^{o} + RT \text{ In } K \quad ...(1)$$

Substituting $\Delta G = -nFE$ and $\Delta G^{\circ} = -nFE^{\circ}$ into equation (1) we have,

$$-nFE = -nFE^{\circ} + RT \text{ In } K$$

Divide both sides of the equation by – nF we have,

$$E = E^{\circ} - RT/nF \text{ In } K \quad ...(2)$$

When In is converted to log then equation (2) can be rewritten as,

$$E = E^{\circ} - 2.303\ RT/nF \log K \quad ...(3)$$

At standard temperature T = 298K, the R is 8.314, F is 96,500 coulombs which is equal to 0.0592 V, so equation (3) turns into,

$$E = E^{\circ} - 0.0592/n \log K.$$

The equation above indicates that the electrical potential of a cell depends upon the reaction quotient K of the reaction. As the redox reaction proceeds reactants are consumed thus concentration of reactants decreases. Conversely the products concentration increases due to the increased products formation. As this happens the cell potential gradually decreases until the reaction is at equilibrium at which $\Delta G = 0$.

Consider the following redox reaction,

$$Mn + ne \rightleftharpoons M.$$

For such a redox reversible reaction the free energy change (ΔG) and its equilibrium constant (K) are interrelated as,

$$\Delta G = -RTl\ n\ K + RTln\frac{[\text{Product}]}{[\text{Reactant}]}$$

$$= \Delta G^\circ + RTl\ n\frac{[\text{Product}]}{[\text{Reactant}]} \quad ...(1)$$

Where,

$\Delta G^\circ =$ Standard free energy change. The equation (1) is known as Van't Hoff isotherm.

The decrease in free energy (−ΔG) in the above reaction will produce electrical energy. In the cell if the reaction involves transfer of 'n' number of electrons the 'n' Faraday of electricity will flow. If E is the emf of the cell, then the total electrical energy (nEF) produced in the cell is,

$$-\Delta G = nEF$$

Or,

$$-\Delta G^\circ = nEF \quad ...(2)$$

Where,

$-\Delta G =$ Decrease in free energy change.

$-\Delta G^\circ =$ Decrease in standard free energy change.

Comparing equation (1) and (2) it becomes,

$$-nEF = -nE^\circ F + RT\ In\frac{[M]}{[M^{n+}]} \quad ...(3)$$

Dividing the above equation (3) by – nF,

$$E = E^\circ - \frac{RT}{nF}l\ n\frac{[M]}{[M^{n+}]}$$

(Activity of solid metal [M] = 1)

$$E = E^\circ - \frac{RT}{nF}l\ n\frac{[M]}{[M^{n+}]}$$

$$= E^{\circ} + \frac{RT}{nF} l\, n\left[M^{n+}\right]$$

$$E = E^{n} - \frac{RT}{nF} l\, n \frac{[Product]}{[Reactant]}$$

$$E = E^{\circ} - \frac{RT}{nF} .2.303 \log\left[M^{n-}\right] \qquad ...(4)$$

When,

$$R = 8.314\ JK^{-1}\ mole^{-1}$$

F = 96.500 Coulomb

T = 298 k

$$E = E^{\circ}_{red} + \frac{0.0591}{n} \log\left[M^{af}\right] \qquad ...(5)$$

Similarly for oxidation potential,

$$E = E^{\circ}_{cell} + \frac{0.0591}{n} \log\left[M^{af}\right] \qquad ...(6)$$

The above equation (5) and (6) are known as "Nernst equation for single electrode potential."

Significance

1. Calculate the EMF of the cell.

$$Zn(s) \mid Zn^{2+}\ (0.024\ M) \parallel Zn^{2+}\ (2.4\ M) \mid Zn(s).$$

Understandably the Z^{2+} ions try to move from the concentrated half-cell to a dilute solution. That driving force gives rise to 0. 0592 V.

At equilibrium concentrations the two half cells will have to be equal in which the voltage will be zero.

2. The equilibrium constant K may be calculated using standard cell potential E^{o} for the reaction.

3. From the standard cell potentials we can calculate the solubility product for the following reaction:

$$A_gCl = A_g^{\ +} + C_{l-1.}$$

3.2 Electro Chemical Series and uses of this Series: Standard Electrodes and Concentration Cells

Electrochemical Series

The electrode potentials of various electrodes are arranged in the order of increasing standard reduction potentials with respect to hydrogen scale. This arrangement is known as electrochemical series.

Importance of electrochemical series:

1. Calculation of Standard EMF of a Cell

The standard EMF of a cell can be calculated using the standard reduction potentials of right hand side and left hand side electrodes from the emf series.

$$E^o_{cell} = E^o_R - E^o_L.$$

EMF

"EMF of a cell is defined as the algebraic difference between the reduction potentials of the cathode and the anode."

The measured EMF of a cell is,

$$E_{cell} = E_{cathode} - E_{anode}$$

$$E_{cell} = E_{right} - E_{left}$$

Where $E_{cathode}\left(E_{right}\right)$ and $E_{anode}\left(E_{left}\right)$ are the reduction electrode potentials of the cathode and anode respectively.

2. Relative ease of Oxidation and Reduction

Higher the value of standard reduction potential greater will be the tendency to get reduced. For example in a electrochemical series fluorine has high positive value of reduction potential (+2.87 V) and is reduced. Lower the value of standard reduction potential greater is the tendency to get oxidized. In a electrochemical series lithium has the lower negative reduction potential value (-3.05V) and is oxidized.

3. Predicting Spontaneity of a Reaction

The spontaneity of a reaction can be determined from the standard emf of the cell. If the emf of a cell is positive, the reaction is spontaneous. If it is negative the reaction is non-spontaneous.

$$E^{o}_{cell} = E^{o}_{r} - E^{o}_{l} = +ve \text{ (spontaneous)}.$$

$$E^{o}_{cell} = E^{o}_{r} - E^{o}_{l} = -ve \text{ (non spontaneous)}.$$

4. Displacement Behaviour of Hydrogen

If zinc electrode is immersed in H_2SO_4 solution it displaces the hydrogen from acid solution.

$$Z_n + H_2SO_4 \rightarrow Z_nSO_4 + H_2^-$$

If silver electrode is immersed in H_2SO_4 solution it will not displace the hydrogen from acid solution because it has higher reduction potential value.

$$Ag + H_2SO_4 \rightarrow \text{No reaction}$$

Use of EMF Series

- Calculation of standard emf of the cell: E° Cell = E° R.H.E - E°L.H.E.
- Relative ease of oxidation or reduction: Higher the positive value of E° greater is the tendency to get reduced whereas higher is negative value of E° greater is tendency to get oxidized.
- Displacement of one element by the other Metal which lies higher in the series can displace those elements which lie below them in the series.
- Determination of equilibrium constant for the reaction:

  $$-\Delta G^{o} = RT \ln K = 2.303\ RT \log K$$

  $$\log K = -\nabla G^{o}/2.303\ RT$$

Since,

$$-\nabla G^{o} = nFE^{o}$$

$$\log K = nFE^{o}/2.303RT$$

5. Hydrogen Displacement Behavior

Metal placed above H_2 in the emf series will displace the hydrogen from an acid solution.

$$Zn + H_2SO_4 \rightarrow Z_nSO_4 + H_2$$

$E^\circ Zn = -0.76V$

$Ag + H_2SO_4 \rightarrow$ No reaction

$E^\circ Ag = +0.80V$

Predicting spontaneity of a reaction: If E° is positive the cell reaction is spontaneous and if E° is negative the cell reaction is not feasible.

3.2.1 Standard Electrodes- Hydrogen Electrode and Calomel Electrode

Standard Electrodes

It is a tendency of a metal to loose or gain electron when it is immersed in its own solution at 25°C and 1M Concentration.

Example: $Zn / ZnSo_4(1M)$ at 25°C.

When we build a cell to measure a potential of a chemical reaction, each pole of the cell contains an electrode that transfers electrons from the solution to the wire or vice versa.

There are two main types of electrodes:

- Indicator electrodes - Electrodes that respond to an analyte concentration.
- Reference electrodes - Electrodes that maintain a fixed potential.

An indicator electrode may be as simple as a piece of Pt wire that is inert, so it does not take part in a reaction but simply provides a path for the electrons to flow.

Many times an indicator electrode may be made of a metal that takes part in the ½ reaction so indicator electrodes can be simple or more complicated.

A best example is the Standard Hydrogen Electrode.

Standard Hydrogen Electrode

This is a great reference electrode because we know its potential is 0 so there is no built in potential that we have to compensate for.

There are two widely used alternative electrodes that are much safer to use.

Silver-Silver Chloride electrode:

$$AgCl(s) + e^- \text{ ------> } Ag(s) + Cl^-$$

The potential of this electrode is based on the ½ reaction and it has potentials of +.222 V if we use 1M KCL or .197 V is used for saturated KCL. Since both Ag and AgCl are solids the only thing that will vary voltage of this cell is Cl hence the two different potentials. The saturated KCL is more reproducible in the lab because we just add KCl to the electrode until we see crystals and we don't need to get the concentrations exactly as 1M.

The silver-Silver Chloride is slightly expensive to make, so there is a less expensive alternative.

The Calomel Electrode

The potential for this electrode is governed by the equation:

$$1/2Hg_2Cl_2(s) + e^- \text{-------->} Hg(l) + Cl^-$$

Again all but Cl are solids so we get 2 potentials +.268 if KCl = 1M +.241 if KCl is saturated.

Given a potential measured against any one of these three common reference electrodes we should be able to convert the potential to measure against one of the other electrodes.

For example we measured a new potential gains of an SHE electrode and found it has a +.2V potential then the potential we measured would be against a Saturated Calomel Electrode.

Our measured potential is +.2 relative to 0 and the SHE is at +.241 from 0, so using this electrode we would measure a +.041 (.241-.2) potential.

We measure a potential of +.1 relative to a SHE electrode then the measure relative to a Ag/AgCl electrode is,

.241 + .1 = .341

.341-.197 = .144

Measurement of Electrode Potential using Calomel Electrode (CE)

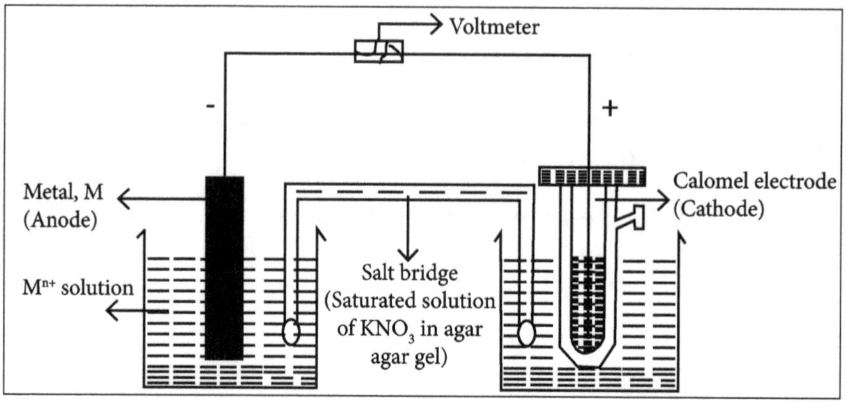

Measurement of electrode potential using calomel electrode.

In order to measure the potential of a given electrode using calomel electrode, a galvanic cell is obtained by combining both the electrodes according to cell conventions.

If a given electrode undergoes oxidation w.r.t. CE, then it is taken as anode and CE is taken as cathode.

Galvanic cell obtained in this case is represented as,

$$M \mid M^{n+} \parallel KCl \mid Hg_2Cl_2(s) \mid Hg(1)$$

At anode, oxidation occurs, $M \rightarrow M^{n+} + ne^-$

At cathode, reduction occurs, $Hg_2Cl_2 + 2e^- \rightarrow 2\,Hg + 2Cl^-$

EMF of cell,

$$E_{Cell} = E_{cathode} - E_{Anode}$$

$$E_{Cell} = ECE - EM^{n+}/M$$

$$E_{Mn} + /M = ECE - E_{cell}$$

If a given electrode undergoes reduction w.r.t. CE, then it is taken as cathode and CE is taken as anode.

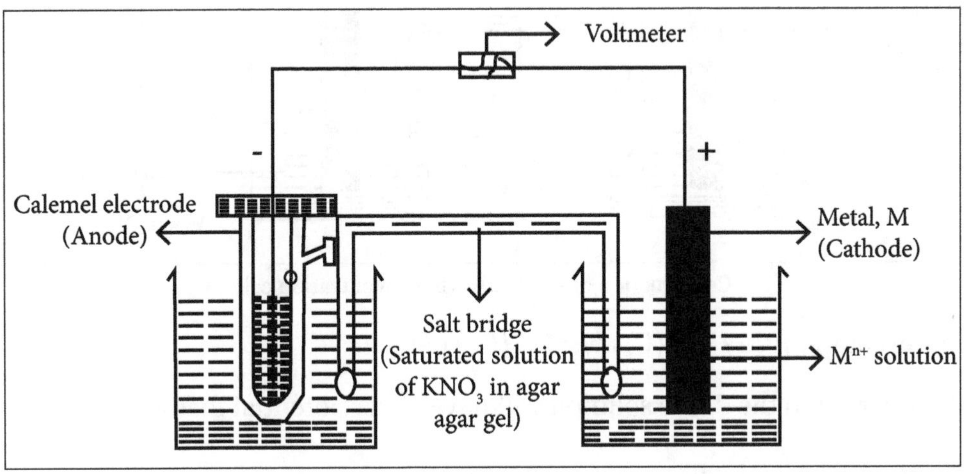

Galvanic cell obtained in this case is represented as; $Hg_{(l)} \mid Hg_2Cl_{2(s)}) \mid KCl \parallel M^{+n} \mid M$

At anode, oxidation occurs, $2\,Hg + 2Cl^- \rightarrow Hg_2Cl_2 + 2^{e-}$

At cathode, reduction occurs, $M^{n+} + n^{e-} \rightarrow M$

EMF of cell, $E_{Cell} = E_{cathode} - E_{Anode}$

$$E_{Cell} = E_{Mn} + /M - ECE$$

$$E_{Mn} + /M = ECE + E_{cell}$$

3.2.2 Concentration Cells

Concentration cell is a type of galvanic cell in which electrode & electrolyte, present in both the half cells are same but only the concentration of metal or electrolyte is different.

Two types of concentration cells:

- Electrolyte concentration cell.
- Electrode concentration cell.

Electrolyte concentration cells: It is a type of galvanic cell in which electrode & electrolyte present in both the half cells are same but only the concentration of electrolyte is different.

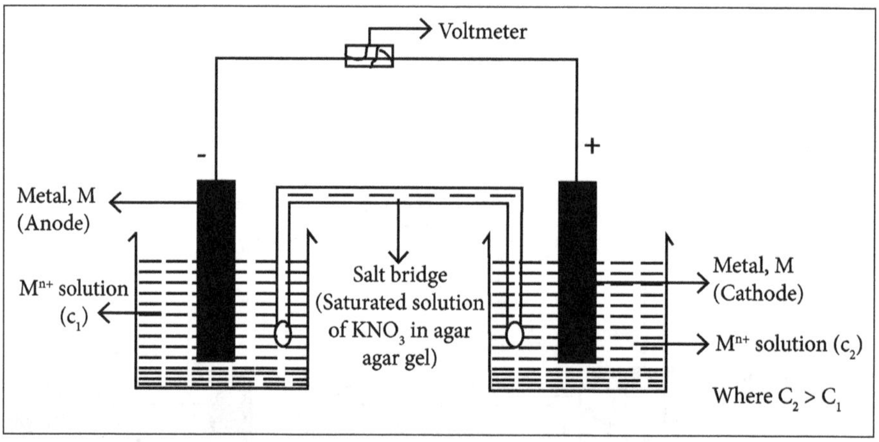

Construction of an electrolyte concentration cell.

Cell representation: $M/M^{n+}(C_1)//M^{n+}(C_2)/M$

Metal immersed in the dilute solution will have lower potential that act as anode,

$$M \rightarrow M^{n+}(C_1) + ne^-$$

Metal immersed in the concentrated solution will have higher potential which act as cathode,

$$M^{n+}(C_2) + ne^- \rightarrow M$$

Net cell reaction is: $M^{n+}(C_2) \rightarrow M^{n+}(C_1)$

Nernst equation for electrolyte concentration cell is,

$$E_{Cell} = E^{o}_{Cell} + 0.0591/n \log \left[M^{n+}\right]cathode/\left[M^{n+}\right]Anode$$

$$E^{o}_{Cell} = E^{o}_{cathode} - E^{o}_{anode} = 0$$

$$E_{Cell} = 0.0591/n \log \left[M^{n+}\right]cathode/\left[M^{n+}\right] Anode$$

$$E_{Cell} = 0.0591/n \log \left(C_2 / C1\right) \text{ at } 298k$$

E_{Cell} is $+ve$ and reaction is spontaneous only when $C_2 > C_{1.}$

Problems

1. Let us write the electrode reactions and calculate the EMF of the given cell at 298K, Ag(s)AgNO$_3$ (0.018M)) AgNO$_3$ (1.2M)Ag(s))).

Solution:

Given:

$$Ag_{(s)}AgNO_3 \ (0.018M))\ AgNO_3 \ (1.2M)Ag_{(s)}))$$

Formula to be used:

$$E_{cell} = \frac{0.0591}{n}\log\left(\frac{C_2}{C_1}\right)$$

At anode: $Ag_{(s)} \rightarrow Ag^{+} + e^{-}$

At cathode: $Ag^{+} + e^{-} \rightarrow Ag_{(s)}$

$$\text{W.K.T } E_{cell}\frac{0.0591}{n}\log\left(\frac{C_2}{C_1}\right) \text{ at } 298K$$

$$E_{cell} = 0.0591 \log\left(\frac{1.2}{0.018}\right)(n = 1)$$

$$E_{cell} = 0.1078 \text{ V.}$$

2. Let us calculate the emf of copper concentration cell at 25° C, where the copper ions ratio in the cell is 10.

Solution:

Given:

Ratio of copper ions = 10

Formula to be used:

$$\frac{[Cu^{+2}]_{cathode}}{[Cu^{+2}]_{anode}} = \frac{C_2}{C_1}$$

$$E_{cell} = \frac{0.0591}{n}\log\left(\frac{C_2}{C_1}\right)$$

$$\frac{[Cu^{+2}]_{cathode}}{[Cu^{+2}]_{anode}} = \frac{C_2}{C_1} = 10$$

W.K.T $E_{cell} = \frac{0.0591}{n}\log\left(\frac{C_2}{C_1}\right)$; at 298 K

$$E_{cell} = \frac{0.0591}{2}\log(10)$$

$$E_{cell} = 0.0296\text{ V.}$$

3. A cell contains two hydrogen electrodes. The negative electrode is in contact with a solution of 10-6 M hydrogen ions. The emf of the cell is 0.118 volt at 25°C. Let us calculate the concentration of hydrogen ions at the positive electrode.

Solution:

Given:

Emf of the cell is 0.118 volt at 25°C

Formula to be used:

$$E_{cell} = 0.0591/2\log\left([H^+]_{cathode}\right)/[10^{-6}]^2$$

The cell may be represented as,

$$Pt|H_2(1\text{ atm})|H^+||H^+|H_2(1\text{ atm})|Pt$$

Anode: $H_2 \rightarrow 2H^+ + 2e^-$

Cathode: $2H^+ + 2e \rightarrow H_2$

$$E_{cell} = 0.0591/2 \log\left(\left[H^+\right]_{cathode}\right)/\left[10^{-6}\right]^2$$

$$0.118 = (0.0591) \log \left(\left[H^+\right]_{cathode}\right)/10^{-6}$$

$$\log\left[H^+\right]_{cathode}/10^{-6} = 0.118/0.0591 = 2$$

$$\left[H^+\right]_{cathode}/10^{-6} = 10^2$$

$$\left[H^+\right]_{cathode} = 1^{0-6} = 10^{-4}\ M$$

4. The emf of the cell Ag|Agl is 0.05 MK/Sol. NH4NO3|10.05 M AgNO3|Ag is 0.788 volt at 25°C. The activity co-efficient of KI and silver nitrate in the above solution is 0.90 each. Calculate (i) the solubility product of Agl and (ii) the solubility of Agl in pure water at 25°C.

Solution:

Given:

Emf of the cell,

Ag|Agl is 0.05 MK/Sol

$AgNO_3 \mid Ag$ is 0.788 volt at 25°C

Co-efficient of KI and silver nitrate is 0.90

Formula to be used:

Solubility product of $Agl = \left[Ag^{+]}\left[I^-\right]$

Solubility of Agl = √(Solubility product of Agl)

Ag+ ion concentration on $AgNo_3$ side = 0.9 × 0.5 = 0.045 M

Similarly I- ion concentration in 0.05 M KI solution = 0.05 × 0.9 - 0.045 M

$$E_{cell} = 0.0591/1 \log\left[Ag^+\right](R.H.S.)/\left[Ag^+\right](L.H.S.) = 0.0591 \log 0.045/\left[Ag^+\right](L.H.S.)$$

Or,

$$\log\ 0.045/\left[Ag^+\right](L.H.S.) = 0.788/0.0591 = 13.33$$

$$[Ag^+]\text{L.H.S.} = 0.045/(2.138 \times 1013) = 2.105 \times 10^{-15}\ \text{M}$$

Solubility product of $AgI = [Ag^{+]}[I^-]$

$$= 2.105 \times 10^{-15} \times 0.045$$

$$= 9.427 \times 10-17$$

Solubility of AgI = √(Solubility product of AgI)

$$= \sqrt{(9.472 \times 10^{-17})}$$

$$= 9.732 \times 10^{-9}\ \text{g mol L}^{-1}$$

$$= 9.732 \times 10^{-9} \times 143.5\ \text{g L}-1$$

$$= 1.396 \times 10^{-6}\ \text{g L}^{-1}$$

3.3 Batteries-Dry Cell, Ni-Cd Cells, Ni-Metal Hydride Cells, Li Cells and Zinc-Air Cells

Battery

A battery is a device that consists of two or more galvanic cell connected in series or parallel or via both, which converts the chemical energy into electrical energy through the redox reaction.

Example: Lithium ion battery, Lead acid battery, Nickel-Cadmium battery, etc.

Uses

- Batteries are used in watches, calculators and pacemakers for heart hearing aids, computers, emergency lightning in hospitals, car engines, military and space applications.

Basic Components of Battery

The battery consists of four major components:

- Anode: It releases electrons into the external circuit by undergoing oxidation.

  $M \rightarrow Mn^+ + e^-$

- Cathode: It accepts the electrons coming from anode through external circuit.

  $Mn^{+} + e^{-} \rightarrow M$

- Electrolyte: It provides the medium for transfer of ions inside the cell between the anode and cathode. A solution of an acid, alkali or salt having high ionic conductivity is commonly used as an electrolyte.

- Separator: It is used to separate anode and cathode compartments in a battery to prevent internal short circuiting. It allows the ions from anode and cathode. Ex: Cellulose, nafion membranes.

Batteries

It is a device which converts the chemical energy into electrical energy and vice versa.

Example: Lead-Acid battery and NICAD.

Emf Variation with Size

The terminal voltage of a battery cell depends on the chemicals and materials used in its construction and not on its physical size.

Batteries are Classified into Primary and Secondary Battery

1. Primary Battery or Primary Cells

A battery which cannot be recharged, cell reactions are irreversible and discarded when the battery has delivered all its electrical energy.

Example: Dry cell or $Zn-MnO_2$ cell, $Li-Mno_2$ cell.

2. Secondary Battery

A battery which can be recharged, cell reactions are reversible and a battery which after discharging, can be recharged.

Example: Lead storage cell and nickel cadmium cell.

3. Reserve Batteries

In reserve batteries, one of the components is stored separately and is incorporated into the battery when required.

Example: Mg-AgCl and Mg-CuCl battery.

Characteristics of a Battery

- Voltage (cell potential).
- Current.
- Capacity.
- Electricity storage density.
- Energy efficiency.
- Cycle life.
- Shelf life.

1. Voltage or Cell Potential (EMF)

The voltage of a battery is given by the equation,

$$E_{cell} = (EC - EA) - \eta A - \eta C - iR_{cell}$$

Where EC and EA are the Electrode potentials of cathode and anode respectively, ηA and ηC are the over potentials at cathode and anode respectively and iR_{cell} is the internal resistance of the cell.

Conditions to Derive Maximum Voltage from a Battery

- Potential difference must be high.
- Over potentials at cathode and anode should be minimum.
- Internal resistance of the cell must be low.

2. Current

For efficient discharge, electron should flow at uniform rate in the electrolyte current is a measure of rate of flow of electrons during discharge. It is the amount of charge flowing per unit and is expressed in ampere per second. Batteries provide direct current.

According to Ohms law, current $I=V/R$. At higher resistance more potential difference is required to force the current through the cell.

3. Capacity

Battery capacity is a measure of the charge stored by a battery; its corresponding Si unit is Ampere-hour (Ah) and the amount of electrical energy the battery that delivers over certain period. Capacity depends on the size of the battery.

The theoretical capacity may be calculated using faradays relation,

$$C = wnF/M,$$

Where C is the capacity

w is the mass of the active material

M is its molar mass

n is the number of moles electrons.

4. Electricity Storage Density

It's a measure of the charge per unit mass stored in the battery. The mass of a battery includes masses of electrolyte current collectors, terminals and other subsidiary elements. Lighter subsidiary elements lead to high storage density. Example: 7g of lithium anode gives 96500 C whereas, for the same charge, 65g of zinc would be required.

5. Energy Efficiency

Is defined as the ratio of useful energy output to the total energy input. A battery should have high energy efficiency.

$$\%\ \text{Energy efficiency} = \frac{\text{Energy released during discharge} \times 100}{\text{Energy required during recharge}}$$

6. Cycle Life

Cycle life of a battery is the number of charge-discharge cycle that can be achieved before failure occurs. The greater is the average depth of discharge, the shorter the cycle life. It is applicable only for the secondary batteries.

It is necessary that during charging the active material is regenerated in a suitable state for discharge. The discharge-charge cycle depends on the chemical composition and distribution of active materials in the cell.

7. Shelf Life

Some of the batteries can be stored for many years. The duration of the storage of a cell without self-discharge or loss of performance is referred as shelf-life.

Dry Cell

Recently the most popular dry cell battery to be used has been the alkaline-cell battery. In the zinc-carbon battery, the zinc is not easily dissolved in basic solutions. Though it is fairly cheap to construct a zinc-carbon battery, the alkaline-cell battery is favored because it can last much longer. Instead of using the NH_4Cl as an electrolyte the alkaline-cell battery will use the NaOH or KOH instead. The same reaction will occur where the zinc is oxidized and it will react with OH^- instead.

$$Zn^{2+} + 2OH^- \rightarrow Zn(OH)^2$$

Once the chemicals in the dry cell battery can no longer react together, then the dry-cell battery is dead and cannot be recharged. Alkaline electrochemical cells have a much longer lifetime but the zinc case still becomes porous as the cell is discharged and the substances inside the cell are still corrosive. Alkaline cells produce 1.54 volts.

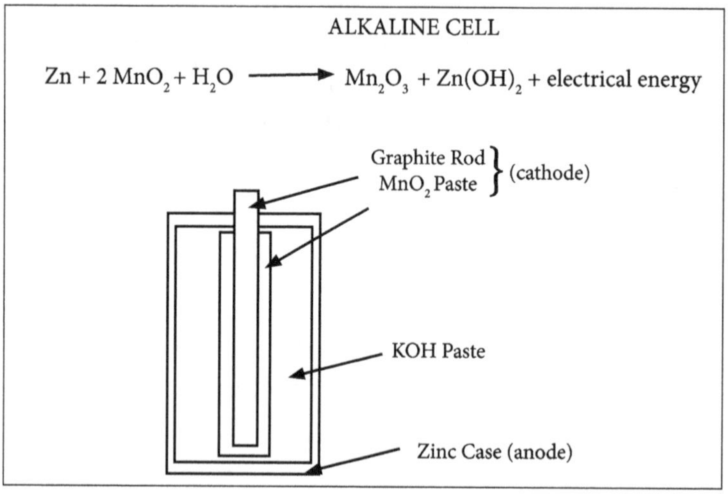

Alkaline cell.

Nickel–Cadmium Battery (Ni-Cd Cells)

A nickel-cadmium storage cell consists of cadmium as anode and NiO_2 paste as cathode and KOH as the electrolyte.

The cell is represented as $Cd\,|Cd(OH)_2\,||KOH(aq)\,|\,NiO_2\,|Ni.$

Construction and Working

When the nickel battery operates, Cd is oxidized to Cd^{2+} ions at anode and the insoluble $Cd(OH)_2$ is formed. NiO_2 is reduced to Ni^{2+} ions which further combines with OH- ions to form $Ni(OH)_2$. It produces about 1.4 V. The following cell reactions occur.

Anodic reaction: $Cd\,(s) + 2OH^- \rightarrow Cd(OH)2(s) + 2e^-$.

Cathodic reaction: $NiO_2(s) + 2H_2O + 2e^- \rightarrow Ni(OH)_2(s) + 2OH^-$.

Overall Cell Reaction During Discharging

$$Cd(s) + NiO_2(s) + 2H_2O \rightarrow Cd(OH)_2(s) + Ni(OH)_2(s) + \text{Energy}.$$

From the above cell reactions, it is clear that $Cd(OH)_2$ and $Ni(OH)_2$ are deposited at both the anodes and cathodes respectively. So this can be reversed by recharging the cell.

Overall Cell Reaction During Recharging

The cell can be charged by passing the electric current in opposite direction. The electrode reactions get reversed. As a result Cd is deposited on the anode and NiO on the cathode.

$$Cd(OH)_2(s) + Ni(OH)_2(s) + \text{Energy} \underset{\text{discharging}}{\overset{\text{charging}}{\rightleftharpoons}} Cd(s) + NiO_2(s) + 2H_2O.$$

Advantages of Ni-Cd Battery

- It has longer life than lead - acid battery.
- It is portable and rechargeable cell.
- It can be easily packed like dry cell since it is smaller and lighter.

Uses

- It is used in gas electronics flash units.
- It is used in calculators.
- It is used in transistors, cordless electronic appliances, etc.

Nickel-Metal Hydride Cell

- Anode — A metal hydride, MH.
- Cathode — Nickel oxy hydroxide/Ni.
- Electrolyte — KOH.

A relatively new technology is adopted in the case of the chargeable sealed nickel-metal hydride battery with characteristics similar to those of the sealed Ni-Cd batteries. The Ni-MH battery uses hydrogen absorbed in a metal alloy for the active negative material whereas the cadmium is used in the Ni-Cd battery and that makes the noticeable difference between the two.

A higher energy density can be achieved in the case of metal hydride electrode than the cadmium electrode. Thus, a smaller amount of the negative electrode is used in the

Ni-metal hydride. This allows for a larger volume for the +ve electrode, which results in the higher capacity or longer service life for the metal hydride battery. Moreover, as the metal hydride battery is free of Cd, it is considered more environmentally friendly than the Ni-Cd battery and may reduce the problems associated with the disposal of rechargeable nickel batteries.

Nickel-metal hydride batteries consist of a positive plate of a highly porous sintered or felt nickel substrate impregnated with the nickel hydroxide as its principal active material, a negative plate of a highly porous structure using a perforated LaNi5 alloy grid. A synthetic non-woven material separates the two electrodes, which serves as a medium for absorbing the electrolyte and sealing the plate provided with a self-resealing safety vent.

The chemistry of the electrode reactions of Ni-MH battery can be described as follows:

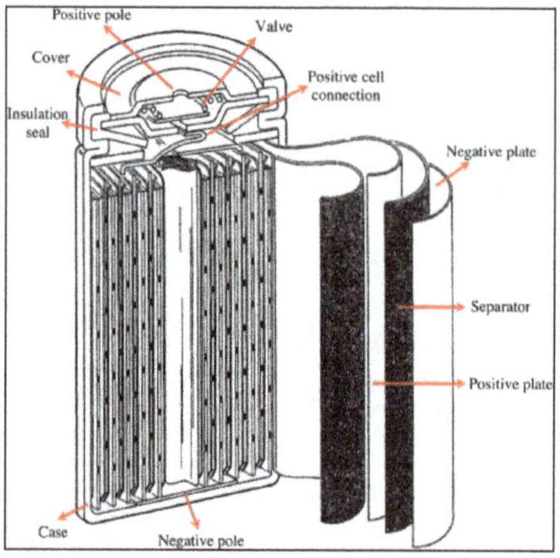

Ni-MH cell.

In charged state of N_i-MH battery, the nickel oxy-hydroxide is the active material of the +ve electrode. This is same as the positive electrode in the Ni-Cd battery.

In the charged state of the Ni-MH battery, hydrogen is stored in a hydrogen absorbing alloy as metal hydride and $LaNi_5$. This metal alloy is capable of undergoing a reversible hydrogen absorbing-desorbing reaction as the battery is charged and discharged.

An aqueous solution of KOH is the major component of the electrolyte, with a minimum amount of electrolyte absorbed by the separator and the electrodes. As can be seen from the overall reaction given below, the chief characteristics of the principle behind a Ni-MH battery is that the hydrogen moves from +ve to -ve electrode during charge and in reverse order during the discharge. With the electrolyte taking no part in the reaction; which means that there is no accompanying increase or decrease in electrolyte.

The discharge electrode reactions of the Ni-MH battery are as follows:

At anode,

$$MH + OH^- \leftrightarrow M + H_2O + e;\ E^o = 0.83\ V$$

At cathode,

The nickel oxyhydroxide is reduced to nickel hydroxide,

$$NiOOH + H_2O + e \leftrightarrow Ni(OH)_2 + OH^-; Eo = 0.52\ V$$

The overall reaction on discharge is,

$$MH + NiOOH \leftrightarrow M + Ni(OH)_2; E^o = 1.35\ V$$

The process is reversed during charge of the Ni-MH battery.

Advantages

The following are the advantages of a Ni-MH battery:

- High capacity.
- No maintenance required.
- Minimum environmental problem.
- Rapid recharging capability.
- Long cycle life.
- Long shelf life in state of charge.

Applications

Nickel-metal hydride batteries are used in computers, cellular phones and in other portable and consumer electronic applications where high specific energy is required:

- Lithium cells: Introduction, construction, working and applications of Li-MnO_2 and Li-ion batteries.
- Lithium is theoretically 'active material for negative electrode' of electrochemical cells owing to its least noble nature and low specific gravity:
- A 'Primary cells' with metallic lithium electrodes and non-aqueous electrolytes were successfully introduced into the market. The outstanding features in comparison with the conventional batteries with aqueous electrolytes are—high voltage, high energy density [i.e., both the volumetric and gravimetric energy densities are high], low self-discharging rate and wide range of operation.
- Thus, the 'secondary batteries' with metallic lithium negative electrodes have attracted much attention as a candidate for the battery with high energy density and much effort has been made in developing secondary lithium batteries.

Many practical problems, however, have been encountered in the development of rechargeable lithium batteries. Among them are:

- Poor cycle life.
- Need for long charging time.
- Poor safety characteristics.

Li-ion Battery

Otherwise referred as Secondary battery.

Construction

- Anode: Lithium atoms intercalated in layered graphite / Carbon metal.
- Cathode: Lithium atoms intercalated in layered structure of MO_2.
- Electrolyte: The electrolyte is lithium salt such as $LiPF_6$ dissolved in organic solvents.
- Separator: Made of Poly propylene membrane.

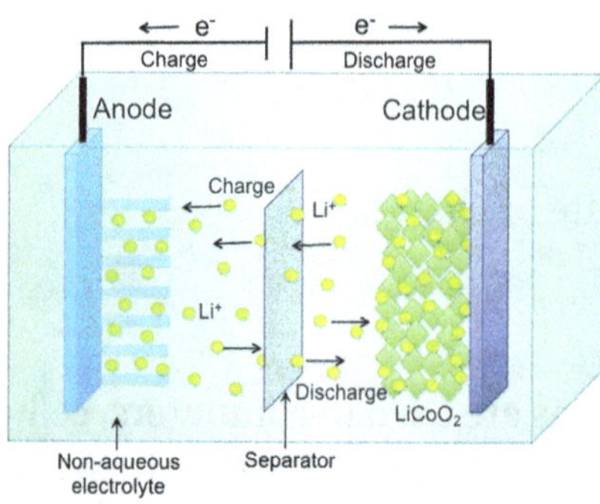

Li-ion battery.

Cell representation: $Li|Li^+,C/LiPF_6$ in organic solvents$|$ $Li-MO_2$.

Electrode reactions are:

During discharging, the lithium atoms present in graphite layer are oxidized, liberating electrons and lithium ions. Electrons flows through the external circuit to cathode and lithium ions flow through the organic electrolyte towards the cathode.

At anode: $Li-C_6 \rightarrow Li^+ + e^- + 6C$

At cathode, the lithium ions are reduced to lithium atoms and are inserted into the layered structure of metal oxide.

At cathode: $MO_2 + Li^+ + e^- \rightarrow Li-MO_2$

During charging, the lithium atoms present in layered structure of metal oxide are oxidized, liberating electrons and lithium ions. Electrons flows through the external circuit and lithium ions flow through the organic electrolyte towards graphite carbon electrode.

At anode: $Li - MO_2 \rightarrow MO_2 + Li^+ + e^-$

At graphite electrode, the lithium ions are reduced to lithium atoms and are inserted in to the layered structure of graphite.

At cathode: $Li^+ + e^- + 6C \rightarrow Li - C6$

Applications: Used in cell phone, Electrical vehicles, Laptops, Aerospace applications, Portable LCD TV, etc.

Zinc-Air Battery

- Anode - Granulated Zn powder.
- Cathode - Air/C.
- Electrolyte - KOH 6M.

The production of electrochemical energy in Zn/air battery is due to the use of oxygen from the atmosphere. The diffused oxygen acts as a cathode reactant in the battery. The air cathode catalytically promotes the reaction of oxygen with an aqueous alkaline electrolyte and is not consumed or changed during the discharge. When an alkaline electrolyte is used in the Zn/air battery, it is necessary to increase only the amount of zinc present to increase battery capacity.

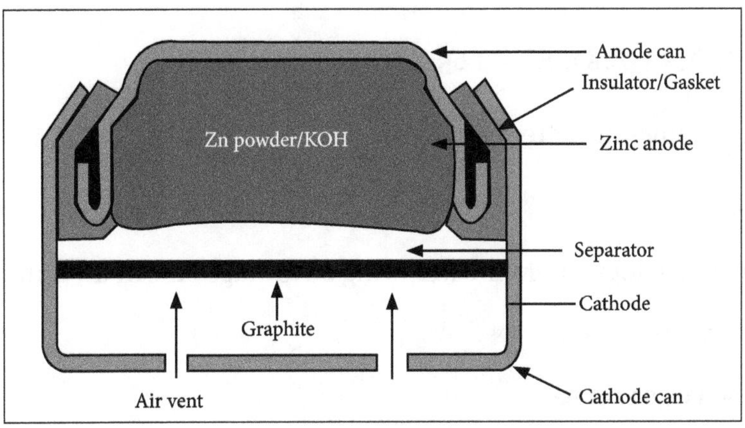

A typical zinc-air cell.

The air cathode acts only as a reaction site and is not consumed. The reason for the increased energy density in the Zn/air cell is because of the larger volume containing the active material. Since the air cathode has infinite life, the electrical capacity of the cell is determined only by the anode capacity, resulting in at least a doubling of energy density.

Theoretically, the air cathode has infinite use of life and its physical size and its electrochemical properties remain unchanged during cell discharge. A schematic representation of a typical Zn/air cell is shown in the above figure. A loose granulated powder of Zn mixed with electrolyte [KOH] acts as the zinc anode material and in some cases, a gelling agent is used to mobilize the composite and ensure adequate electrolyte contact with zinc granules. The outer metal acts as the cathode of the battery and a plastic gasket insulates the anode active materials and the cathode as shown in the above figure.

The chemistry of the electrode reactions taking place in zinc/air battery are as follows:

At anode,

$$Zn \rightarrow Zn^{2+} + 2e; E^{o} = 1.20\ V$$

$$Zn^{2+} + 2OH^{-} \rightarrow Zn\left(OH\right)2$$

$$Zn\left(OH\right)_2 \rightarrow ZNO + H_2O$$

At cathode,

$$½O_2 + H_2O + 2e \rightarrow 2OH^{-}; E^{o} = 0.45\ V$$

The overall efficiency of zinc/air battery,

$$Zn + ½\ O_2 \leftrightarrow ZnO;\ E^{o} = 1.65\ V$$

The output of the zinc-air battery is 1.65 V.

Advantages

Zn/air battery technology offers the following advantages for many applications:

- High energy density.
- Flat discharge voltage.
- Long shelf life.
- No ecological problems.
- Low cost.

- Capacity independent of load and temperature.

Applications

Zn-air batteries have been most successfully employed as a power source for hearing aids. Other applications include the voice transmitters, electronic pagers, portable battery chargers, various medical devices and so on.

3.4 Corrosion

Corrosion is the deterioration of a metal as a result of chemical reactions between it and the surrounding environment.

$$\text{Metal} \underset{\text{Extraction}}{\overset{\text{Corrosion}}{\longleftrightarrow}} \text{Metal ore}$$

Both the type of metal and the environmental conditions, particularly what gases that are in contact with the metal, determine the form and rate of deterioration.

A Corrosion is an undesirable process. Due to this corrosion there is limitation of progress in many areas. As a result of corrosion, the cost of replacement of materials and equipments is unlimited.

Metals and alloys are used as fabrication or construction materials in engineering. If the metals or alloy structures are not properly maintained, then they deteriorate slowly by the action of moisture, atmospheric gases and other chemicals. This phenomenon of destruction of metals and alloys is referred as corrosion.

Corrosion of metals is defined as a spontaneous destruction of metals in the course of their chemical, electrochemical or biochemical interactions with the environment. Thus, it is exactly the reverse of extraction of metals from ores.

Example: Rusting of iron

A layer of reddish scale and powder of oxide (Fe_3O_4) is formed on the surface of iron metal.

A green film of basic carbonate $[CuCO_3 + Cu(OH)_2]$ is formed on the surface of copper, when it is exposed to moist-air containing the carbon dioxide.

Consequences

- Efficiency of the machine is lost due to corrosion products.

- Corroded equipment must be replaced frequently.
- Products get contaminated due to released toxic products.
- Necessity of over designing.
- Failure of plants.

Causes of Corrosion

In nature, metals occur in two different forms:

- Native state.
- Combined state.

1. Native State

The metals exist as such in the earth crust then the metals are present in native state.

Native state means free or uncombined state. These metals are non-reactive in nature. They are noble metals which have very good corrosion resistance. Example: Au, Pt, Ag, etc.

2. Combined State

Except the noble metals, all other metals are highly reactive in nature which undergoes reaction with their environment to form stable compounds called ores and minerals. This is the combined state of metals. Example: Fe_2O_3, ZnO, C_aCO_3, PbS etc.

Metallic Corrosion

The metals are extracted from their metallic compounds. During the extraction, the ores are reduced to their metallic states by applying energy in the form of various processes.

In a pure metallic state, the metals are unstable as they are considered in excited state. Therefore as soon as the metals are extracted from their ores the reverse process begins and form metallic compounds which are thermodynamically stable (i.e., lower energy state).

Hence, when the metals are used in various forms, they are exposed to environment and when heated, the exposed metal surface begin to decay (conversion to more stable compound). This is the basic reason for metallic corrosion.

$$\text{Metal} \underset{\text{Metallurgy-Reduction}}{\overset{\text{Corrosion-Oxidation}}{\rightleftharpoons}} \text{Metallic Compund} + \text{Energy}$$

Although corroded metal is thermodynamically more stable than pure metal but due to corrosion, the useful properties of a metal like malleability, hardness, ductility, luster and electrical conductivity are lost.

Effects of Corrosion

The economic and social consequences of corrosion includes:

- Due to formation of corrosion product over the machinery, the efficiency of the machine gets failed and leads to plant shut down.
- The corroded equipment must be replaced.
- The products contamination or loss of products occurs due to corrosion.
- Corrosion releases the toxic products.
- Preventive maintenance like metallic coating or organic coating is required.
- Health (e. g., from pollution due to a corrosion product or due to the escaping chemical from a corroded equipment).

3.4.1 Theories of Corrosion (Chemical and Electrochemical)

Classification or theories of corrosion

Based on the environment, corrosion is classified into:

- Dry or chemical corrosion.
- Wet or electrochemical corrosion.

1. Dry or Chemical Corrosion

This type of corrosion is due to the direct chemical attack on metal surfaces by the atmospheric gases such as oxygen, hydrogen sulphide, halogen, sulphur dioxide, nitrogen or anhydrous inorganic liquid, etc. The chemical corrosion is defined as the direct chemical attack on metals by the atmospheric gases present in the environment.

Example:

- Silver materials undergo chemical corrosion by Lower case atmospheric H_2S gas.
- Iron metals undergo chemical corrosion by HCl gas.

Types of dry or chemical corrosion:

- Corrosion by oxygen or oxidation corrosion.
- Corrosion by hydrogen.
- Liquid metal corrosion.

1. Corrosion by Oxygen or Oxidation Corrosion

Oxidation corrosion is brought about by the direct attack of oxygen at low or high temperature on metal surfaces in the absence of moisture. Alkali metals (Li, k, Na, etc.,) and alkaline earth metals (Mg, Sn, Ca, etc.,) are rapidly oxidized at low temperature. At high temperature, almost all metals (except Au, Ag and Pt) are oxidized. The reactions of oxidation corrosion are as follows:

Mechanism

i. Oxidation takes place at the surface of the metal forming metal ions.

$$M \rightarrow M_2^{+} + 2e^{-}$$

ii. Oxygen is converted to oxide ion (O^{2-}) due to the transfer of electrons from metal.

$$n/2\,O_2 + 2ne^{-} \rightarrow n\,O_2^{-}$$

iii. The overall reaction of the oxide ion reacts with the metal ions to form metal oxide film.

$$2\,M + n/2O_2 \rightarrow 2\,Mn^{+} + nO_2^{-}$$

The Lower case Nature or the Lower case Oxide formed plays an important part in oxidation corrosion process.

Metal + Oxygen → Metal oxide (corrosion product)

When oxidation starts, a thin layer of oxide is formed on the metal surface and the nature of this film decides the further action.

2. Corrosion by Other Gases (by Hydrogen)

i. Hydrogen Embrittlement

Loss in ductility of a material in the presence of hydrogen is referred as hydrogen embrittlement.

Mechanism

This type of corrosion occurs when a metal is exposed to hydrogen environment. Iron liberates the atomic hydrogen with hydrogen sulphide in the following way,

$$Fe + H_2S \rightarrow F_eS + 2H$$

Hydrogen diffuses into the metal matrix in this atomic form and gets collected in the voids present inside the metal. Further, diffusion of the atomic hydrogen makes them combine with each other and forms hydrogen gas,

$$H + H \rightarrow H_2 \uparrow$$

Collection of these gases in the voids develops very high pressure, causing cracking or blistering of metal.

2. Decarburization

The presence of carbon in steel gives sufficient strength to it. But when the steel is exposed to hydrogen environment at high temperature, atomic hydrogen is formed,

$$H_2 \xrightarrow{\text{Heat}} 2H$$

Atomic hydrogen reacts with the carbon of the steel and produces the methane gas,

$$C + 4H \rightarrow CH_4$$

Hence the carbon content in steel is decreased. The process of decreasing the carbon content in steel is known as decarburization.

Collection of methane gas in the voids of steel develops high pressure which causes cracking. Thus, the steel loses its strength.

3. Liquid Metal Corrosion

This is due to chemical action of flowing liquid metal at high temperatures on solid metal or alloy. Such corrosion occurs in devices used for nuclear power. The corrosion reaction involves either,

- Dissolution of a solid metal by a liquid metal.

Or,

- Internal penetration of the liquid metal into the solid metal.

Both these modes of corrosion cause weakening of the solid metal.

ii. Wet or Electrochemical Corrosion

Electrochemical corrosion involves:

- The formation of anodic and cathodic areas or parts in contact with each other.
- Presence of a conducting medium.

- Corrosion of anodic areas.
- Formation of corrosion product somewhere between the anodic and cathodic areas.

This involves flow of electron-current between the anodic and cathodic areas.

At anodic area, the oxidation reaction takes place, so the anodic metal is destroyed by either dissolving or assuming combined state (such as oxide, etc). Hence the corrosion always occurs at anodic areas.

$$M\ (\text{metal}) \rightarrow M^{n+} + n\,e^{-}$$

$$M^{n+}\ (\text{metalion}) \rightarrow \text{Dissolves in solution}$$

→ Forms compounds such as oxide

At cathodic area, the reduction reaction takes place (gain of electrons), usually cathode reactions do not affect the cathode since most of the metals cannot be further reduced. So at cathodic part the dissolved constituents in the conducting medium accepts the electrons to form some ions like OH and $O_2^{\,-}$.

Cathodic reaction consumes the electrons by:

- Evolution of hydrogen.
- Absorption of oxygen depending on the nature of the corrosive environment.

(a) Hydrogen Evolution Type

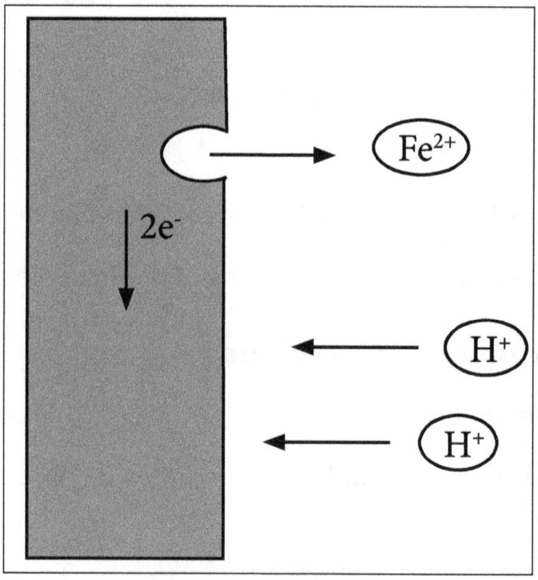

Hydrogen evolution type.

All metals above hydrogen in the electrochemical series have a tendency to get dissolved in acidic solution with simultaneous evolution of hydrogen.

It occurs in acidic environment. Consider the example of iron.

At anode: $Fe \rightarrow Fe^{2+} + 2e^-$

These electrons flow through the metal from anode to cathode where H^+ ions of acidic solution are eliminated as hydrogen gas.

At cathode: $2\,H^+ + 2\,e^- \rightarrow H_2 \uparrow$

The overall reaction is: $Fe + 2H^+ \rightarrow Fe_2^+ + H_2$

(b) Oxygen Absorption Type

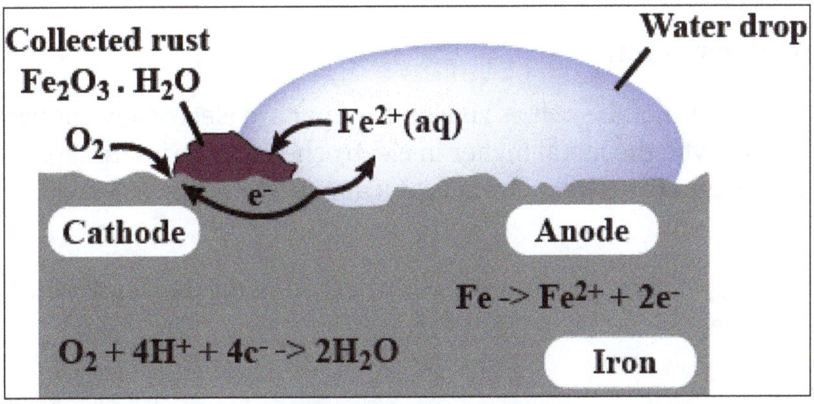

Oxygen absorption type.

Rusting of iron in a neutral aqueous solution of electrolytes in the presence of atmospheric oxygen is a common example of this type of corrosion. The surface of iron is usually coated with a thin film of iron oxide. However, if this iron oxide film develops some cracks, anodic areas are created on the surface, while the well metal parts acts as cathodes.

At Anode: Metal dissolves as ferrous ions with liberation of electrons,

$$Fe \rightarrow Fe^{2+} + 2e^-$$

Cathode: The liberated electrons are intercepted by the dissolved oxygen,

$$\frac{1}{2}\,O_2 + H_2O + 2\,e^- \rightarrow 2OH$$

The Fe^{2+} ions and OH^- ions diffuse and when it meets ferrous hydroxide it is precipitated,

$$Fe^{2+} + 2OH^- \rightarrow Fe(OH)_2$$

(i) If enough oxygen is present, ferrous hydroxide is easily oxidized to ferric hydroxide.

$$4Fe(OH)_2 + O_2 + 2H_2O \rightarrow 4Fe(OH)^3 \ (\text{Yellow rust } Fe_2O_3.H_2O)$$

(ii) If the supply of oxygen is limited, the corrosion product may be even black anhydrous magnetite Fe_3O_4.

Types of Electrochemical Corrosion

The electrochemical corrosion is classified into the following two types:

- Galvanic corrosion.
- Differential aeration or concentration cell corrosion.

1. Galvanic Corrosion

When two dissimilar metals such as zinc and copper are electrically connected and exposed to an electrolyte, the metal higher in electrochemical series undergoes corrosion. In this process, the more active metal acts as a anode whereas, the less active metal acts as cathode.

In the above example, zinc acts as anode, is attacked and gets dissolved whereas, copper acts as cathode.

Mechanism

In acidic solution, corrosion occurs by the hydrogen evolution process whereas, in neutral or slightly alkaline solution, oxygen absorption occurs. The electron-current flows from the anode metal zinc to the cathode metal copper.

$$Zn \rightarrow Zn^{2+} + 2e^- \ (\text{Oxidation})$$

Thus, it is evident that, the corrosion occurs at the anode metal, whereas, the cathodic part is protected from the attack.

Example:

- Steel screws in a brass marine hardware.
- Lead-antimony solder around copper wire.
- A steel propeller shaft in bronze bearing.
- Steel pipe connected to copper plumbing.

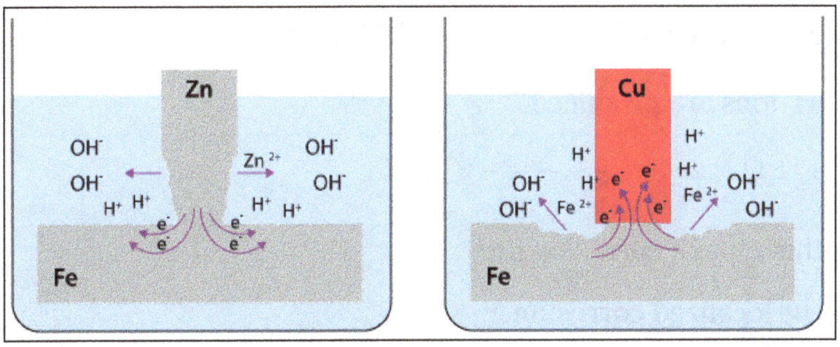

Galvanic corrosion.

2. Concentration Cell Corrosion

It is due to electrochemical attack on the metal surface, exposed to an electrolyte of varying concentrations or of varying aeration. It occurs when one part of the metal is exposed to a different air concentration from the other part. This causes a difference in potential between differently aerated areas. It has been found experimentally that poor oxygenated parts are anodic.

Example:

- The metal part immersed in water or in a conducting liquid is called water line corrosion.
- The metal part is partially buried in soil.

Explanation

If a metal is partially immersed in a conducting solution the metal part above the solution is more aerated and becomes cathodic. The metal part inside the solution is less aerated and thus, becomes anodic and suffers corrosion.

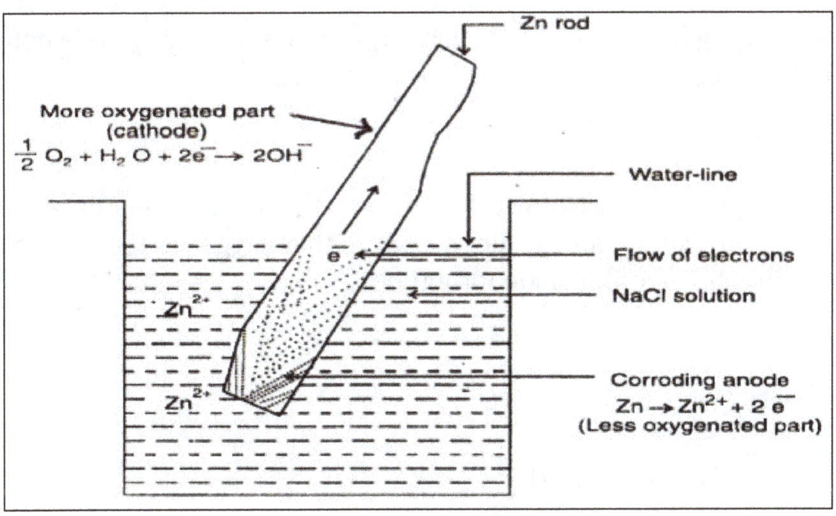

Concentration cell corrosion.

At anode: Corrosion occurs $M \text{ -------> } M^{2+} + 2e^-$

At Cathode: OH^- ions are produced.

$$\frac{1}{2} O_2 + H_2O + 2e^- \text{ ---------> } 2OH.$$

Examples for this type of corrosion are:

- Pitting or localized corrosion.
- Crevice corrosion.
- Pipeline corrosion.
- Corrosion on wire fence.

3.5 Formation of Galvanic Cells by Different Metals, by Concentration Cells, by Differential Aeration and Waterline Corrosion

Formation of Galvanic Cells by Different Metals

It occurs when two dissimilar metals are in contact with each other in a corrosive conductive medium; a potential difference is setup resulting in a galvanic current. The two metals differ in their tendencies to undergo oxidation. The metal with lower electrode potential or the more active metal acts as anode whereas, the metal with higher electrode potential acts as cathode. The potential difference is the main factor for corrosion to take place. The anodic metal undergoes corrosion whereas, cathodic metal gets unaffected.

Example: When iron is in contact with copper, iron has lower electrode potential, acts as anode and undergoes oxidation. It can be written as,

$$Fe \rightarrow Fe^{2+} + 2e^-$$

Copper having higher electrode potential acts as cathode and is unaffected. Depending on the corrosive environment near the cathode, either oxygen is absorbed or hydrogen is evolved resulting in hydroxide ion formation.

At cathode,

$$O_2 + 2H_2O + 4e^- \rightarrow 4OH^-$$

$$2Fe^{2+} + 4OH^- \rightarrow 2Fe(OH)_2$$

$$4Fe(OH)_2 + O_2 + 2H_2O \rightarrow 2[Fe_2O_3 \cdot 3H_2O]$$

Yellow Rust

The rate of galvanic corrosion depends upon the potential difference between anodic and cathodic metals, ratio of anodic and cathodic area and environmental factors and tendency of the metal to undergo passivity etc.

Other examples:

When Fe is in contact with Sn then Fe acts as anode and Sn acts as cathode but when Fe is in contact with Zn, Fe acts as cathode whereas, Zn acts as anode.

Formation of Galvanic Cells by Concentration Cells

A concentration cell is an electrochemical cell in which both half cells are of the same type, but with different electrolyte concentrations. The following cell notations are examples of concentration cells.

$$Cu|Cu^{2+}(aq,\ 0.0010\ M)||Cu^{2+}(aq,\ 1.0\ M)|Cu$$

$$Ag|Ag^{+}(aq,\ 0.0010\ M)||Ag^{+}(aq,\ 0.10\ M)|Ag$$

In concentration cells, the half-cell with low electrolyte concentration acts as an anode half-cell and one with the higher electrolyte concentration acts as the cathode half-cell. At the anode half-cell, a oxidation reaction occurs to increase the electrolyte concentration and at the cathode half-cell, a reduction reaction occurs to decrease its electrolyte concentration. Oxidation-reduction reactions will continue until the electrolyte concentrations in both half cells becomes equal.

At anode half cell: $Cu(s) \rightarrow Cu^{2+}(aq) + 2e^-, (\text{in } 0.0010\ M\ Cu^{2+})$

At cathode half cell: $Cu^{2+}(aq) + 2e^- \rightarrow Cu(s),\ (\text{in } 0.50\ M\ Cu^{2+})$

Formation of Galvanic Cells by Differential Aeration Corrosion

Differential aeration corrosion occurs when metal surface is exposed to the differential air or oxygen concentration. This develops galvanic cell on the metal and initiates corrosion. The part of the metal exposed to the lower oxygen concentration acts as an anode and the part of the metal exposed to higher concentration acts as the cathode, so that poorly oxygenated region undergoes corrosion.

Example:

When a metal strip of iron is partially immersed in an aerated solution of NaCl the concentration of O_2 is higher at the top of the surface than inside the solution. Since cathodic reaction requires oxygen, the cathodic area tends to concentrate near the waterline so that bottom portion of the specimen acts as anode where corrosion starts.

At anode: $Fe \rightarrow Fe_2 ++ 2e^-$

At cathode: $O_2 + 2H_2O + 4e^- \rightarrow 4OH^-$

Types

There are two types:

- Waterline corrosion.
- Pitting corrosion.

Formation of Galvanic Cells by Waterline Corrosion

This is a case of differential aeration corrosion generally observed in steel water tanks, ocean going ships etc., in which, the portion of the metal is always under water. The part of the metal below the waterline is exposed only to dissolved oxygen whereas, the part above the waterline is exposed to higher concentration of atmospheric oxygen.

Thus, the metal part below the waterline acts as anode and undergoes corrosion whereas, the metal part above the waterline, which is more oxygenated is unaffected and acts as cathode. This type of corrosion is generally observed in ships floating in seawater for a long period of time.

3.6 Passivity of Metals, Pitting Corrosion and Galvanic Series

Passivity of Metals

Passivity is defined as a condition of corrosion resistance due to the formation of thin surface film under oxidizing conditions, some metals and alloys having simple barrier films with reduced corrosion of active potential. For instance, in the presence of concentrated fuming nitric acid, iron is inert despite the highly oxidizing conditions of the solution and when the acid is diluted with water iron remains inert, while initially corroding vigorously evolving brown, nitrous oxide gas when the surface is lightly scratched. Passivity, is displayed by chromium, aluminum, iron, nickel, titanium and many of their alloys. This passive behaviour results from the formation of a highly adherent and very thin oxide film on the metal surface, which acts as a protective barrier to further corrosion.

Stainless steels are highly resistant to corrosion in a rather wide variety of atmospheres as a result of passivation. Chromium is noted for formation of very stable, thin resistant surface in less oxidizing conditions when alloys with other metals especially iron which is having minimum 12% Cr is passive in most aerated solutions.

Pitting Corrosion

The mechanism of pitting corrosion involves setting up of differential aeration or concentration cell. Metal area covered by a drop of water, dust, sand, scale etc., is the aeration or concentration cell.

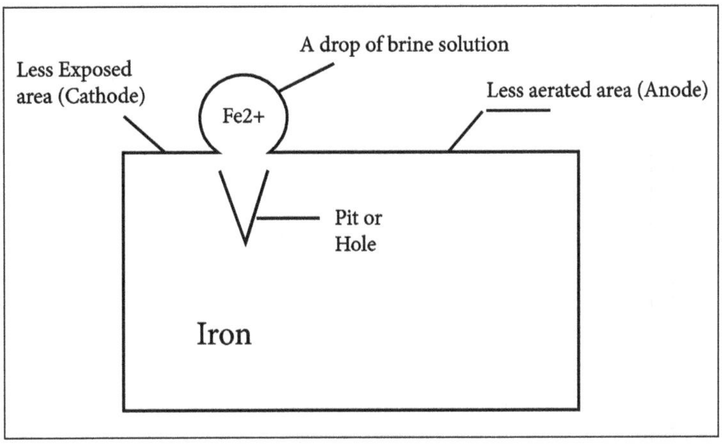

Pitting corrosion.

Pitting corrosion is explained by considering a drop of water or brine solution on a metal surface. The area covered by the drop of salt solution has less oxygen and acts as anode. This area suffers corrosion, the uncovered area acts as cathode due to high oxygen content.

It has been found that, the rate of corrosion will be more when the area of cathode is larger and when the area of the anode is smaller. Hence, there is more material around the small anodic area and results in the formation hole or pit.

At anode: Fe is oxidized to Fe^{2+} and releases electrons.

$$Fe^{-} \rightarrow Fe^{2+} + 2e^{-}$$

At cathode: Oxygen is converted to hydroxide ion.

$$\frac{1}{2} O_2 + H_2O + 2e^{-} \rightarrow 2OH^{-}$$

The net reaction is, $Fe + 2OH^{-} \rightarrow Fe(OH)_2$

The above mechanisms can be confirmed by using ferroxyl indicator. Since OH- ions are formed at the cathode, this area imparts pink colour with phenolphthalein indicator.

At the anode, iron is oxidized to Fe^{2+} which combines with ferricyanide and shows blue colour.

Galvanic series

An arrangement of alloys and metals in the order of their corrosion resistance in the given environment is known as galvanic series.

Platinum — Higher electrode potential/less reactive and less susceptible for corrosion ↑
Gold
Silver
Titanium
Lead
Copper
Nickel
Stainless steel
Iron
Steel
Tin
Zine
Magnesium — Lower electrode potential/more reactive and more susceptible for corrosion ↓

Different metals have different tendencies to react or corrode. The familiar electrochemical series lists various elements in the order of their standard electrode potentials, the electrochemical series has however, limited utility in corrosion studies and a poor guide in predicting corrosion behavior of metals and alloys in real environment since, it do not take passivity of the metal into consideration.

Example:

Chromium is less noble than iron in ECS, but Cr is a passive metal and its alloys with iron import excellent corrosion resistance. This is contrary to the prediction by ECS. In order to overcome these limitations, galvanic series was introduced. Then metals and alloys are arranged in the order of their corrosion resistance.

Difference between ECS & GS

Electrochemical series	Galvanic series
1. Eo values are measured by dipping metal in their salt solution.	Developed by studying corrosion of metal or alloy in the given environment.
2. Position of the given metal is fixed.	The position may shift.

3. No information regarding position of alloys.	Alloys are included in the series.
4. It predicts relative displacement tendency.	It predicts relative corrosion tendency.

3.6.1 Factors which Influence the Rate of Corrosion

The factors that affect the rate of corrosion are:

1. Factors Connected with the Metal

i. Position of the Metal in the EMF Series

The type of impurity present in it and its electropositive nature decides the corrosion of a metal. For instance, when iron has impurities like tin, copper, etc., iron corrodes since iron is more electropositive than metals like copper and tin. On the other hand when iron is coupled with zinc, zinc corrodes as zinc is more electropositive than iron.

ii. Purity of the Metal

Generally pure metals does not corrode as there is no cathode spot available to induce corrosion.

iii. Surface of the Metal

A rough surface corrodes readily as it collects more dirt and provides more cathode spot for corrosion. A polished surface does not corrode easily.

iv. Stress Corrosion

Stress in a metal surface is produced by mechanical workings such as bending and riveting, improper heat treatment, quenching, pressing, etc. The portion subjected to more stress acts as anode and the other portion act as cathode. This leads to the formation of stress corrosion. Stress corrosion is noted in fabricated articles of certain alloys like high zinc brasses and nickel brasses.

Caustic embrittlement noted in boilers is a typical example for stress corrosion which is due to the attack of alkali present in water on stressed boiler metal.

v. Anode to Cathode Area Ratio

When a bigger cathode area covers a smaller anode area severe corrosion is noted in the anode spot. This is known as erosion. It is frequently encountered in piping agitators, condenser tubes etc., where turbulent flow of gases and vapours remove the coated surfaces resulting in differential cells. Removal of surface coatings can also be caused by striking or rubbing activities of solids on the coated surfaces.

vi. Physical State of the Metal

The rate of corrosion is influenced by orientation of crystals, grain size, stress, etc. The smaller the grains size of the metal, greater the rate of corrosion.

2. Factors Connected with the Environment

- Provides space for all numbers.
- Nature of the atmosphere.
- Temperature of the atmosphere.
- pH of the atmosphere.
- Amount of moisture in the atmosphere.
- Amount of oxygen in the atmosphere.
- Amount of chemical fumes in the atmosphere.

Examples:

- Provides space for all numbers, buried pipelines and cables passing from one type of soil to another suffer corrosion due to differential aeration.
- Lead pipelines passing through clay and then through sand. Lead pipeline passing through clay get corroded because it is less aerated than sand.
    - In some cases, the corroded product sticks to the surface and absorbs more moisture. This induces further corrosion.

      Example: Rusting of iron, as rust formed over iron absorbs more moisture, rate of rusting of iron increases.
    - In some cases, the corroded product acts as the protective coating which prevents further corrosion.

      Example: Aluminium oxide formed over the surface of aluminium prevents further corrosion and act as a protective coating. This is the basic principle of anodization.
    - In some cases, the corroded product falls out of position exposing the fresh metal surface for further corrosion.

      Example: Magnesium oxide formed over the surface of magnesium falls out of position exposing a fresh surface for further corrosion.

3.6.2 Protection from Corrosion

As the corrosion process is very harmful and losses incurred are tremendous, it becomes necessary to minimize or control corrosion of metals. Corrosion can be stopped completely only under ideal conditions. But the attainment of ideal conditions is not possible. However, it is possible only to minimize corrosion considerably. Since the types of corrosion are so numerous and the conditions under which corrosion occurs are so different, diverse methods are used to control corrosion. As the corrosion is a reaction between the metal or alloy and the environment, any method of corrosion control must be aimed at either modifying the metal or the environment.

Corrosion Inhibitors

Corrosion inhibitor is a substance which reduces the corrosion of a metal, when it is added to the corrosive environment.

Types of Inhibitors

Inhibitors are classified into three types. They are:

- Anodic inhibitors.
- Cathodic inhibitors.
- Vapour phase inhibitors.

1. Anodic Inhibitors

Example: Chromates, Nitrates, Phosphates.

Anodic inhibitors are those that prevent the corrosion reaction occurring at the anode by forming an insoluble compound with the newly produced metal ions. These precipitates are adsorbed on the anode surface forming a protective film and reducing the corrosion rate. Though this type of control is effective it may be dangerous. Since severe local attack can occur, if some areas are uncovered.

2. Cathodic Inhibitors

In an electrochemical corrosion, the cathodic reactions can be of two types, based on the environment.

i. In an acidic solution:

Example: Amines, heavy metal soaps, mercaptans.

In an acidic solutions, the cathodic reaction is due to the evolution of hydrogen.

$$\text{(i.e.) } 2H + 2e^- \rightarrow H_2$$

ii. In a neutral solution:

Example: Sodium sulphite (Na_2SO_3), Hydrazine (N_2H_4). In a neutral solution, the cathodic reaction is,

$$H_2O + \frac{1}{2}O_2 + 2\ e^- \rightarrow 2\ OH^-$$

The corrosion can be reduced in two ways:

- By eliminating oxygen from the neutral solution, thereby formation of OH- ions are inhibited.

This can be done by adding reducing agents like Na_2SO_3, N_2H_4 etc.

- By eliminating the OH- ions from the neutral solution. This can be done by adding Mg, Zn or Ni salts. These react with OH- ions form insoluble hydroxides which are deposited on the cathode forming more or less impermeable self-barriers.

3. Vapour Phase Inhibitors (VPI)

Example: Benzotriazole.

Vapour phase inhibitors are organic inhibitors which readily vapourize and form a protective layer on the metal surface. VPI are used in the protection of storage containers, packing materials, etc.

3.6.3 Design and Material Selection

- Differential metal corrosion can be reduced by avoiding the use of two dissimilar metals. If there use cannot be avoidable, the position of the metals chosen shall be as close as possible in the galvanic series.
- While joining the metals, care should be taken not to leave gaps between them, where some air or liquid can be trapped resulting in differential aeration corrosion.
- Stress corrosion may be avoided by annealing the equipment.
- To reduce the rate of corrosion, anodic material must be as large as possible whereas, the cathodic material should be small.

3.6.4 Cathodic Protection

Cathodic Protection is a technique used to control the corrosion of a metal surface by making it the cathode of an electrochemical cell. A simple method of protection, connects the metal to be protected to a more easily corroded sacrificial metal serves as the anode. The sacrificial metal then corrodes instead of the protected metal. For structures, such as long pipelines where passive galvanic cathodic protection is not adequate an external DC electrical power source is used to provide sufficient current.

Principle

The principle involved in the cathodic protection is to force the metal to act as a cathode.

The important cathodic protections are:

- Sacrificial anodic protection.
- Impressed current cathodic protection.

Sacrificial Anodic Protection Method

In this method, the metallic structure to be protected is made cathode by connecting it with more active metal. So that all the corrosion will concentrate only on the active metal. The artificially made anode thus, gradually gets corroded protecting the original metallic structure. Hence, this process is also known as sacrificial anodic protection. Aluminium, zinc, magnesium are used as sacrificial anodes.

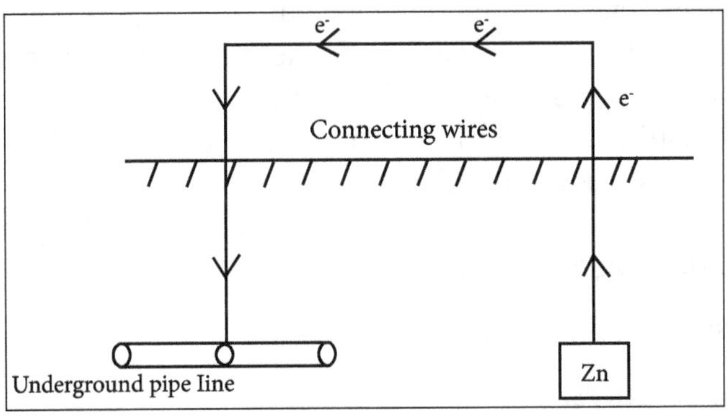

Anode protection.

Impressed Current Cathodic Protection

In this method, an impressed current is applied in the opposite direction of the corrosion current to nullify it and the corroding metal is converted from anode to cathode.

This can be done by connecting the negative terminal of the battery of the metallic structure, to be protected and positive terminal of the battery is connected to an inert anode. Inert anodes used for this purpose are graphite, platinized titanium. The anode is buried in a 'back fill'. The 'back fill' provides good electrical contact to anode.

Applications of Impressed Current Protection

Structures like tanks, pipelines, transmission line towers, underground water pipe lines, ships, etc. can be protected by this method.

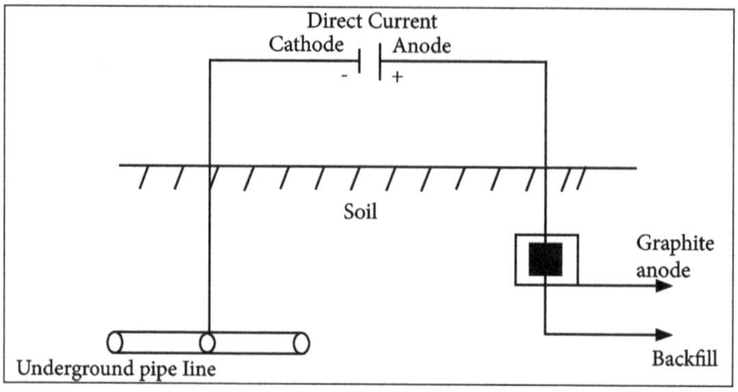

Impressed current protection.

3.7 Protective Coatings: Surface Preparation

Protective Coatings

Corrosion of metal surfaces is a common phenomenon. To protect the metal surfaces from corrosion, the contact between the metal and the corrosive environment is to be cut-off. This is done by coating the surface of the metal with a continues, non-porous material, insert to the corrosive atmosphere. Such a coating is known as protective coating or surface coating. In addition to protective action, such coatings also give a decorative effect and reduce wear and tear.

Objectives of Coating Surfaces

- To enhance wear and scratch resistance.
- To increase hardness.
- To insulate electrically.
- To insulate thermally.
- To impart decorative colour.
- To prevent corrosion.

Surface Preparation

Surface preparation is the essential first stage treatment of a substrate before the application of any coating. The performance of a coating is influenced by its ability to adhere properly to the substrate material. Correct surface preparation is the most important factor affecting the total performance of surface treatment. The presence of even small amounts of surface contaminants, such as oil, grease, oxides etc., can physically impair and reduce coating adhesion to the substrate. Chemical contaminants that are not visible, such as chlorides and sulphates, attract moisture through coating systems resulting in premature failure.

The adhesion of coating materials such as zinc by hot dip galvanizing is particularly good due to the metallurgical alloying with the steel surface. In this process, the molten zinc reacts with the steel to form a series of iron or zinc alloy layers of varying composition to form an intimate bond with the substrate.

Alternatively, metal coatings applied by thermal spraying requires a coarse surface profile to maximize the coating bond which is principally by mechanical keying. Organic paint coatings adhere to the surface mainly by polar adhesion which is aided by mechanical adhesion which is especially important for thick coating films. For all types of coatings, the surface condition of the substrate is critical in terms of durability and coating performance.

3.7.1 Metallic (Cathodic and Anodic) Coatings

Metal Coating

Corrosion of metal can be controlled by isolating them from the corrosive atmosphere. This can be done by covering the metal with a layer of another metal. This process is known as metal coating.

1. Anodic Metal Coating (Sacrificial Coating)

It is produced by coating a base metal with more active or more anodic metal to the base metal. For example, iron is coated with Zn, Mg, Al etc.

One of the important properties of this type of coating is that, even if the coating is ruptured, the base metal does not undergo corrosion. The exposed part of the base metal is cathodic with respect to the coating metal and the coating metal only undergoes corrosion thereby, protecting the base metal. The protection exists as long as coating exists. Galvanization is one of the best examples in anodic coating.

Galvanization

It is a process of coating a base metal with zinc metal. This process is usually carried out by hot dipping method.

- The base metal surface is washed properly with organic solvents to remove any organic matter such as oil, grease etc., on the surface.
- It is washed with dilute H_2SO_4 to remove any inorganic matter like rust.
- Finally the base metal is washed well with water and is then air dried.
- The base metal is then dipped in a bath of molten zinc maintained at 425-430°C

and covered with a flux of NH_4Cl to prevent the oxidation of molten zinc. Then excess zinc on the surface is removed by passing a pair of hot rollers so that a proper thin coating is obtained.

Applications

Galvanized articles are mainly used in roofing sheets, fencing wire, buckets, bolts nuts, pipes, tubes, etc., but galvanized articles are not used for preparing and storing food stuffs, since zinc dissolves in dilute acids and become toxic.

2. Cathodic Metal Coating

These are the coatings produced by coating a base metal with more cathodic material. For example, iron is coated with tin, nickel and Cu, but these coatings provides protection only when it is undamaged and absolutely free from gaps, otherwise rapid corrosion of the base metal takes place as a result of formation of large cathodic and small anodic area. Tinning is the best example to explain cathodic coating.

Tinning

It is a process of coating base metal with tin. It can be carried out by hot dipping method:

- The iron sheets are first washed with organic solvents to remove any organic impurities.
- Then it is treated with dilute H_2SO_4, to remove rust.
- Finally it is washed well with water and air dried.
- It is then passed through $ZnCl_2$ flux, so that the molten tin can adhere properly to the metal surface.
- Then base metal is passed through the tank that contains molten tin.
- Finally they are passed through a series of rollers immersed in palm oil. So that uniform, undamaged, continuous deposit of tin takes place. Tinning will provide complete protection against corrosion if it covers the surface completely.

Applications

- Tinned articles are largely used in the manufacturing of containers used for storing foodstuffs.
- Copper utensils are coated with tin so that contamination of food with copper can be prevented.

3.7.2 Methods of Application on Metals (Galvanizing, Tinning, Electroplating and Electroless Plating)

The metal which is coated upon is known as the base metal. The metal applied as coating is referred to as coat metal.

The different methods used for metal coating are:

- Hot dipping:
    - Galvanization
    - Tinning
- Metal spraying
- Cladding
- Cementation:
    - Sherardizing → Cementation with zinc powder is known as sherardizing.
    - Chromizing → Cementation with 55% chromium powder and 45% alumina is known as chromizing.
    - Calorizing → Cementation with aluminium and alumina powder is termed as calorizing.
- Electroplating or electrodeposition
- Electroless plating

Galvanization

The process of coating a layer of zinc on iron is known as galvanizing. The iron article is first cleaned with dilute sulphuric acid to remove traces of rust, dust, etc., at 60-90°c for about 15 -20 minutes. Then, this metal is dipped in a molten zinc bath which is maintained at 430°C.

The surface of the bath is covered with ammonium chloride flux to prevent oxide formation on the surface of molten zinc. The coated base metal is then passed through rollers to correct the thickness of the film. It is used to protect roofing sheets, wires, pipes, tanks, nails, screws, etc.

Tinning

The coating of tin on iron is known as tinning or tin plating. In tinning, the base metal is first cleaned with dilute sulphuric acid to remove surface impurities. It is then passed through molten tin covered with zinc chloride flux. The tin coated article is passed through a series of rollers immersed in palm oil bath to remove the excess tin. Tin coated utensils are used for storing food stuffs, pickles, oils, etc.

Galvanizing is preferred to tinning because tin is cathodic to iron, while, zinc is anodic to iron. So, if the protective layer of the tin coating has any cracks, iron will corrode. Similarly, if the protective layer of the zinc coating has any cracks, iron does not get corroded since it is cathodic. Thus, the corrosion products fill up the cracks preventing corrosion.

Metal Finishing

Metal finishing is the process carried out to modify the surface properties of a metal by electrodeposition of a layer of another metal on the substrate.

Techniques

Important techniques of metal finishing are:

- Electroplating.
- Electroless plating.

Electroplating

Gold Electroplating

Gold electroplating is the process of applying a thin layer of gold onto a desired metal, generally that of silver or copper, through the process known as electroplating.

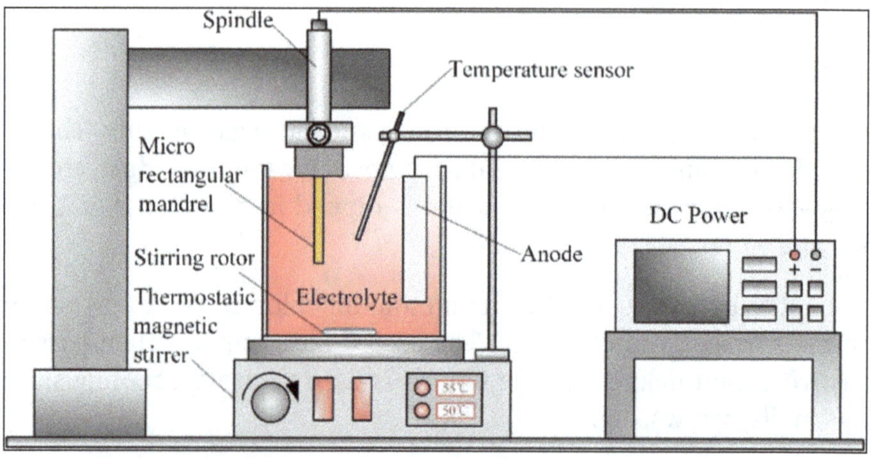

Simple schematic diagram of electroplating process.

Process of Gold Electroplating

When gold electroplating is used on silver in the manufacturing of jewellery, a layer of nickel and copper and must be deposited onto the silver before the electroplating of gold occurs. The reason for this is that if no intermediate layer were provided, the silver atoms would over time, diffuse through the gold plating and cause tarnishing. The nickel and copper layers slows down this process. With electronics, an intermediate layer of nickel is used as copper will, diffuse through the gold layer and cause tarnishing and a sulphide or oxide layer will be formed.

Electroless Plating

Electroless plating is a technique of depositing a noble metal on a catalytically active surface of the metal which is to be protected, by using a suitable reducing agent without the use of electrical energy.

Electroless Plating of Nickel

Principle

The reducing agent reduces the metallic ions to metals which gets plated over the catalytically active surface giving a uniform thin coating.

$$\text{Metal ions} + \text{reducing agent} \rightarrow \text{Metal} + \text{oxidized products}$$

Step 1: Pretreatment and Activation of the Surface

The surface to be plated is first degreased by using organic solvents followed by acid treatment. Activation depends on the type of the metal or alloy or the non-metal used.

Example:

- The surface of the stainless steel is activated by dipping it in hot solution of 50% dilute H_2SO_4.
- The surface of Mg alloy is activated by their coating of Zn and Cu over it.
- Metals and alloys like Al, Cu, Fe, etc., can be directed to Ni-plating without activation.

Step 2: Preparation of Plating Bath

The plating bath consists of the following ingredients:

Nature of the compound	Name of the compound	Quantity (g/l)	Function
1. Coating solution	$NiCl_2$	20	Coating metal
2. Reducing agent	Sodium hypophosphite	20	Metal ions
3. Complexing agent cum exhalant	Sodium succinate	15	Improves quality
4. Buffer	Sodium acetate	10	Controls pH

5.Optimum pH	4.5	-	-
6.Optimum temperature	93°C	-	-

Step 3: Procedure for Plating

The preheated object is immersed in the plating bath for a certain duration, during which the following reduction reaction will occur and the nickel gets coated over the object.

Reactions

At cathode: $Ni^{2+} + 2e^{-} \rightarrow Ni$

At anode: $H_2PO_2^{-} + H_2O \rightarrow H_2PO_3^{-} + 2H^{+} + 2e^{-}$

Net reaction: $Ni_2^{+} + H_2PO_2^{-} + H_2O \rightarrow Ni + H_2PO_3^{-} + 2H^{+}$

Applications

- Electroless Ni-plating is extensively used in electronic appliances.
- Electrodes Ni-plating is used in domestic as well as automobile fields.

4

Chemistry of Advanced Materials

4.1 Nanomaterials: Sol-Gel Method and Chemical Reduction Method of Preparation

Nanomaterials are defined as the materials with at least one external dimension in the size ranging from approximately 1-100 nanometers. Nanoparticles are the objects with all three external dimensions at the nanoscale.

They are either naturally occurring or the incidental byproducts of combustion processes and are physically and chemically heterogeneous, often termed as ultrafine particles.

Salient Properties of Nanomaterials

1. Melting Points

Nanomaterials have a significantly lower melting point and appreciably reduced lattice constants. This is due to huge fraction of surface atoms in the total number of atoms.

2. Optical Properties

Reduction of material dimensions has effects on optical properties. Optical properties of nanomaterials are different from bulk forms. The change in optical properties is caused by two factors:

- The quantum confinement of electrons within the nanoparticles increases the energy level spacing.
- Surface plasma resonance, which is due to smaller size of nanoparticles than the wavelength of incident radiation.

3. Mechanical Properties

Nanomaterials have fewer defects compared to bulk materials, which increases the mechanical strength.

- Mechanical properties of polymeric materials can be increased by the addition of nanofillers.

- As nanomaterials are stronger, harder and more corrosion and wear resistant, they are used in spark plugs.

Sol-Gel Method

The sol-gel process is a wet technique (i.e.) chemical solution deposition technique used for the production of high purity and homogeneous nanomaterials. In solutions, the molecules of nanometer size are dispersed and move around randomly, hence the solution is clear. In colloids, the molecules of size ranging from 20 to 100 are suspended in a solvent.

When mixed with a liquid is known as sol. A suspension that keeps its shape is known as a gel. Thus, the colloids are actually the suspensions of colloids in liquid that keeps up their shape. The formation of sol-gel involves hydrolysis, condensation growth of particles and formation of networks.

Chemical Reduction Method of Preparation

Chemical reduction method is increasingly attractive and provides good potential in terms of accomplishing physicochemical properties, better handling, equipment configuration and the associated cost factors involved in other processes. By utilizing suitable raw materials, reducing agents and stabilizing agents, the nanomaterials of relatively smaller or narrow size distribution and desired shape can be effectively synthesized. Also the stability of the nanomaterials being dispersed into the base material can be established very well without experiencing settlement or agglomeration of the particles.

Chemical reduction method is the frequently used method for the preparation of silver nanoparticles as colloidal and stable dispersions in water or organic solvents occur. Commonly used reducing agents are elemental hydrogen, citrate and borohydride. The reduction of silver ions in aqueous solutions yield colloidal silver with particle diameters of several nanometers. Initially, the reduction of various complexes with Ag^+ ions leads to the formation of silver atoms, which is followed by agglomeration into oligomeric clusters.

These clusters eventually lead to the formation of colloidal silver particles. When the colloidal particles are much smaller than the wavelength of visible light, the solutions have a yellow color with an intense band in the range of 380-400 nm and other less intense or smaller bands at longer wavelengths in the absorption spectrum. This band is attributed to collective excitation of the electron gas in the particles, with a periodic change in electron density at the surface. Controlled synthesis of silver nanoparticles is based on a two-step reduction process. In this technique, a strong reducing agent is used to produce small silver particles, which are enlarged in the secondary step by further reduction with a weaker reducing agent. Chemical reduction of metal salts using various reducing agents in the presence of stabilizer is currently used for the preparation of metal nanoparticles. Reducing agents such as formaldehyde, hydrazine (N_2H_4), sodium borohydride ($NaBH_4$), etc., can be used to reduce a silver containing salt to produce nano-silver particles.

4.1.1 Characterization by BET Method and TEM Method

Characterization by BET Method

The Brunauer, Emmett and Teller (BET) technique is the common method for determining the surface area of powders and porous materials. Nitrogen gas is usually employed as the probe molecule and is exposed to a solid under investigation at liquid nitrogen conditions. The surface area of the solid is determined from the measured monolayer capacity and the knowledge of the cross sectional area of the molecule being used as the probe. In the case of nitrogen, the cross-sectional area is taken as 16.2 $Å^2$ /molecule.

In case of spherical, non-porous particles, the BET surface area is related to the particle radius (R) or diameter (D) and the skeletal density through,

$$BET\left[\frac{m^2}{g}\right] = \frac{4\pi R^2}{\frac{4}{3}\pi R^3 \rho_s} = \frac{3}{\rho_s R} = \frac{6}{\rho_s D}$$

Nanoparticles are being incorporated into a wide range of products including medical devices, nutritional products, drugs, personal care products, plant growth regulators and specialty additives. High surface area systems can be attained either by fabricating small clusters or particles where the surface to volume ratio of each particle is high or by creating materials where the void surface area is high compared to the amount of bulk support material. Small particles such as nanoparticles have a high surface to volume ratio potentially resulting in altered bioavailability, increased rate of dissolution, increased surface reactivity and most importantly, a changed toxicity profile. BET surface area measurements may also be highly relevant when considering the toxicological aspects of novel materials.

Characterization by Transmission Electron Microscope (TEM) Method

TEM is a very powerful tool for material analysis, which helps in revealing the crystal structure, determination of lattice spacing, defects, morphology and size of the particles, etc. TEM analysis reside a high energy beam of electrons shoring through a very thin specimen, resulting in interactions between the electrons and the atoms enabling them to observe particular features.

The principle behind TEM is as same as that of light microscope, but uses electrons instead of light. It can be used to study their chemical composition as well as defects in semiconductors and the growth of layers. High resolution can be used to analyze the density, shape, size and quality of nanoparticles or the nanograins.

4.2 Carbon Nanotubes and Fullerenes: Types, Preparation, Properties and Applications

4.2.1 Carbon Nanotubes

Carbon nanotubes are allotropes of carbon with a nanostructure having a length to diameter ratio greater than 1,00,000. When graphite sheets are rolled into a cylinder, their edges are joined to form carbon nanotubes, (i.e.) carbon nanotubes are extended tubes of rolled graphite sheets.

Nanotube

Nanotubes are one of the widely used materials. It consists of tiny cylinders of carbon and other materials like boron nitride. Nanotubes of carbon and inorganic compound with structures, comparable to the layered structure of graphite have been prepared.

Types of Carbon Nanotubes

Nanomaterials are die materials having components with size less than 100 m, at least in one dimension.

Depending on the way in which the graphite sheets are rolled, two types of CNTs are formed.

- Single-walled nanotubes (SWNTs).
- Multi-walled nanotubes (MWNTs).

1. Single-Walled Nanotubes (SWNTs)

- SWNTs consist of one tube of graphite.
- It is one atom thick having a diameter of 2 nm and a length of 100 μm.
- SWNTs are very important because they exhibit important electrical properties.
- It is an excellent conductor.
- Three kinds of nanotubes are resulted based on the orientation of the hexagon lattice.
    - Arm chair structure: The lines of hexagons are parallel to the axis of the nanotube.
    - Zig zag structure: The lines of carbon bonds are down the center.
    - Chiral nanotubes: It exhibits twists or spirals around the nanotubes.

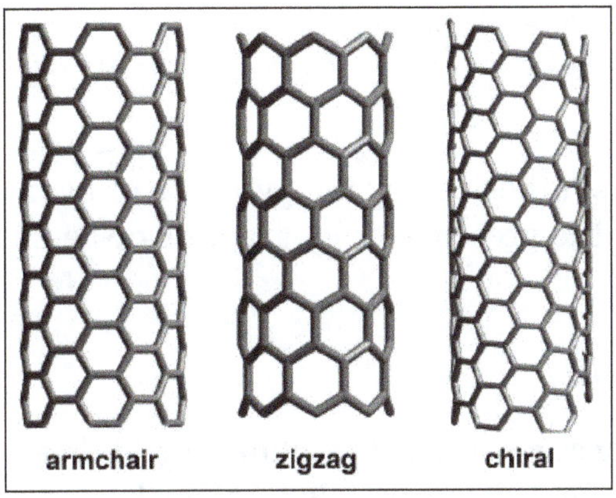

Types of nanotubes.

- It has been confirmed that arm chair carbon nanotubes are metallic while zig zag and chiral nanotubes are semiconducting.

2. Multi-Walled Nanotubes (MWNTS)

- MWNTS also known as nested nanotubes, consist of multiple layers of graphite rolled in on themselves to form a tube.
- It is used for storing fuels such as methane and hydrogen.
- It exhibits both semiconducting and metallic properties.

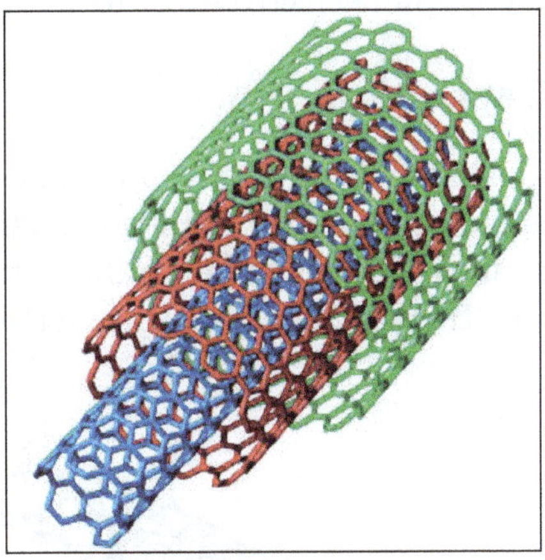

Multi-wall nanotubes (MWNT'S).

Synthesis of CNTS

Carbon nanotubes can be synthesized by any one of the following methods:

- Pyrolysis.
- Laser evaporation.
- Carbons arc method.

- Chemical vapour deposition.

1. Pyrolysis

Carbon nanotubes are synthesized by the pyrolysis of hydrocarbons such as acetylene at about 700°C in the presence of Fe-graphite or Fe-silica catalyst under inert conditions.

2. Laser Evaporation

A quartz tube containing argon gas and graphite target is heated to 1200°C. The quartz tube contains the water cooled copper collector which is present outside the furnace. The graphite target containing small amount of cobalt and nickel acts as a catalyst for the formation of nanotubes.

An intense pulsed laser beam is incident on the target, which results in evaporation of carbon from the graphite. The argon gas is used for the transportation of carbon atoms from the high temperature zone to cooled copper collector, where it condense into nanotubes.

3. Carbon Arc Method

It is carried out by applying direct current and voltage in the range of 60-100A and 20-25V respectively between the graphite electrodes of 10-20 μm diameter.

4. Chemical Vapour Deposition

It involves the decomposition of hydrocarbon or methane gas at 1100°C. As the gas decomposes, carbon atoms are produced. It is then condensed on a cooler substrate which contains catalyst like iron. This method, produces tubes with open end which allows continuous fabrication. It is the most favorable method for scale up and production.

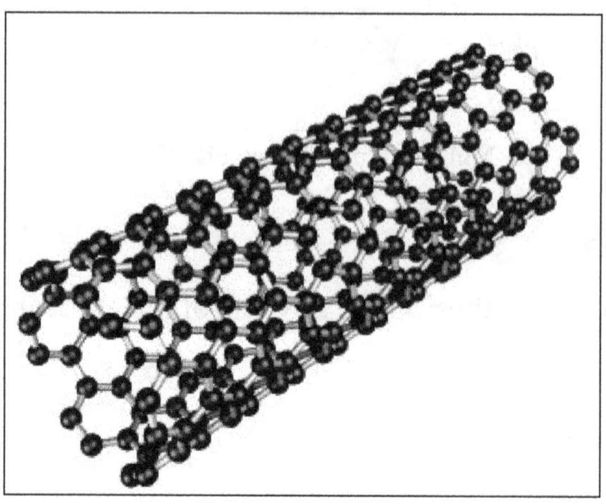

Carbon Nanotubes: Properties and Applications

Properties of Carbon Nanotubes

- CNTs are good electron field emitters.
- CNTs have high electrical and thermal conductivity.
- CNTs have very high tensile strength.
- CNTs are highly flexible, can be bent considerably without any damage.
- CNTs have a low thermal expansion co-efficient.
- CNTs have a high aspect ratio.

Applications of Carbon Nanotubes

CNTs can be used in:

- Air and water filtration systems.
- Biomedical applications.
- Conductive plastics, adhesives and connectors.
- Catalyst supports.
- Ceramic applications.
- Energy storage systems.
- Field emission systems.
- Fibers and fabrics.
- Molecular electronics based on CNTs.
- Thermal conductivity of CNTs.
- Structural composites.
- Thermal materials.

They are also widely used in chemical processing, electronics, energy management and material science.

4.2.2 Carbon Fullerenes

Fullerenes are a family of carbon allotropes, molecules composed entirely of carbon in the form of a ellipsoid, hollow sphere, plane and tube. Fullerenes are similar in structure to graphite, which is composed of stacked sheets of linked hexagonal rings, but may also contain pentagonal or heptagonal rings.

The most familiar carbon fullerene is a molecule with 60 carbon atoms, represented as C_{60}. It was discovered in 1985 and was named as Buckminsterfullerene. It was the first discovered fullerene which possessed 60 atoms of carbon. The name was coined after the American architect Richard Buckminster Fuller.

The C_{60} molecule has a truncated icosahedral structure formed by replacing each vertex on the seams of a football by a carbon atom. There are 20 hexagonal faces and 12 pentagonal faces in a molecule.

The average nearest C - C distance is 0.144 nm, which is very close to that in graphite. Each carbon atom is trigonally bonded to other carbon atoms, same as that in graphite.

Out of the three bonds emanating from each carbon atom, there is one double bond and two single bonds. The hexagonal faces consist of alternating single and double bonds and the pentagonal faces are defined by single bonds. The length of the single bond is 0.146 nm, which is longer than the average bond length (i.e.) 0.144 nm, while the double bonds are shorter, 0.14 nm. The diameter of the C_{60} molecule is 0.71 nm. The inner cavity is capable of holding a variety of atoms.

Types of Fullerene

Fullerenes have many structural variations and they have progressed well. The types of fullerene are given below:

Nanotubes or Cylindrical Fullerenes

They are hollow in shape, having very small dimension. Nanotubes that are made of carbon are generally wide and can range only a few nanometers to several millimeters in length. They have both open and closed ends. It is mainly used in electronics industry, paper batteries and in space technology.

Buckyballs Clusters

It is the smallest fullerene that occurs in nature. Its tiniest member is C_{20} and most common one is C_{60}. The smallest fullerene has great importance in terms of natural occurrence and it can be found in coal or soot.

Mega Tubes

As the name indicates, these tubes have larger diameter than nanotubes. The walls of mega tubes are prepared with different thickness. These types of tubes are generally used to transport of a variety of molecules having different sizes.

Polymers

They are macromolecules connected by covalent chemical bonds. Polymers mainly consists of carbon chains. Under high pressure and high temperature, some two dimensional and three dimensional polymers are formed.

Nano-Onion

It is a form of buckyball solid, having spherical particles.

Linked "ball and chain" Dimers

Two buckyballs linked by a carbon chain.

Fullerene Rings

Preparation of Fullerenes

These are a new form of carbon, discovered in 1985 by the vapourization of graphite under inert atmosphere at low pressure. The properties of fullerenes are different from that of diamond and graphite. Fullerene can be used in electronic devices, superconductors, catalysts, sensors, polymer composites, high energy fuels and medical applications. In graphite, we can find planar layer of carbon atoms, while in fullerenes, we can find tubular layers. The spherical fullerenes are called buckyballs, while the cylindrical ones are called carbon nanotubes or buckytubes. The structure of the fullerene is similar to that of graphite, (i.e.) composed of stacked graphene (an allotrope of carbon, which contains one atom thick planar sheets of densely packed sp^2-bonded carbon atoms. The name graphene is due to its resemblance with graphite and presence of sp^2 carbon atoms) sheets of linked hexagonal rings.

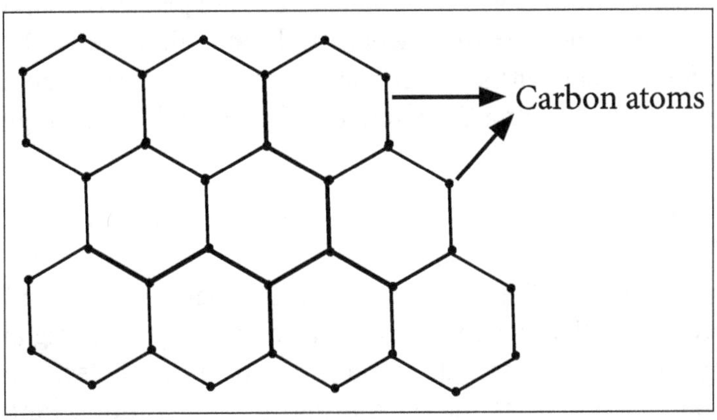

Graphene

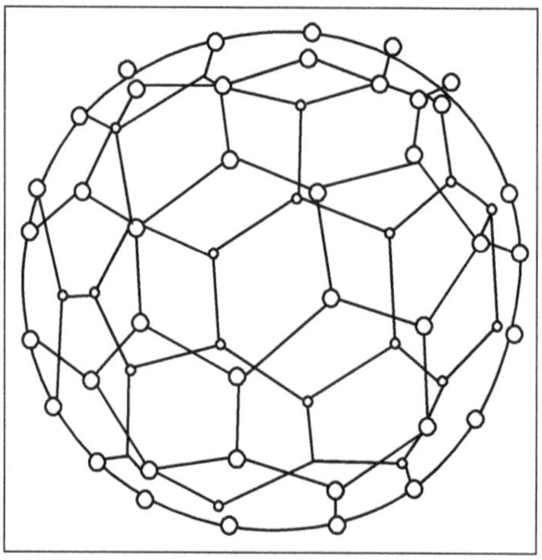

Buckminsterfullerene (C_{60})

Properties

Chemical Properties

The carbon atoms within a fullerene molecule are hybridized with sp^2 and sp^3, of which the sp^2 carbons are responsible for the considerable angle strain presented within the molecule. C_{60} and C_{70} has the capacity to be reversibly reduced with up to six electrons. This high electron affinity results from the presence of triply degenerate low lying LUMOs (lowest unoccupied molecular orbital). Oxidation of the molecule has also been observed still, oxidation is irreversible. C_{60} has a localized pi-electron system, which prevents the molecule from exhibiting super aromaticity properties.

Physical Properties

Fullerenes are extremely strong molecules, capable of resisting great pressures, they will bounce back to their original shape after being subject to over 3,000 atmospheres. Theoretical calculations suggest that a single C_{60} molecule has an effective bulk modulus of 668 GPa when compressed to 75% its size. This property makes fullerenes become harder than diamond and steel, whose bulk moduli are 442 Gpa and 160 GPa, respectively. Fullerenes can withstand collisions of up to 15,000 mph against stainless steel, bouncing back and keeping their shapes. This resembles the high stability of the molecule.

Optical Properties

Delocalized pi-electrons in fullerenes are known to provide exceptionally large non-linear optical responses. Fullerenes finds application in optical limiting and intensity dependent refractive index systems. Additionally, the transfer of electrons from enclosed atoms to the fullerene, enhances the third order non-linear optical effect by orders of magnitude compared to empty cage fullerenes.

Applications

Solar Cells

The high electron affinity and superior ability to transport charge makes fullerenes the best acceptor component. First, they have an energetically deep lying LUMO, which provides the molecule with a very high electron affinity relative to the numerous potential organic donors. The LUMO of C_{60} also allows the molecule to be reversibly reduced with up to six electrons thus, illustrating its ability to stabilize negative charge. A number of conjugated polymer fullerene blends are known to exhibit ultrafast photoinduced charge transfer, with a back transfer that is orders of magnitude slower. The state of the art in the field of organic photovoltaic is currently the Bulk Hetero-junction (BHJ) solar cells based on fullerene derivate [6, 6]-phenyl-C_{61}-butyric acid methyl ester (PCBM), with reproducible efficiencies approaching 5%.

Hydrogen Gas Storage

Due to its unique molecular structure, fullerene is the only form of carbon, which can be potentially and chemically hydrogenated and de-hydrogenated reversibly. When fullerenes are hydrogenated, the C=C double bonds become C-C single bond and C-H bond. The bond strength of single C-C bonds is 83 kcal/mole and theoretical calculations shows that the bond strength of the hydrogenated C-H bond is 68 kcal/mole. This means that for fullerene hydrides, the H-C bond is appreciably weaker that C-C bond. Therefore, when heat is applied to fullerene hydrides, the H-C bond will break before the C-C bond and the fullerene structure should be preserved.

Fullerene Strengthening/Hardening of Metals

Fullerenes offer unique opportunities to harden alloys and metals without seriously compromising their ambient temperature ductility. This is due to the unique characteristics of fullerenes, namely their high reactivity and small size, which enable the dispersion strengthening of metallic matrices with carbide particles that result from interactions between fullerenes and metals.

Fullerene as Precursor to Diamond

Fullerenes have proven to be an excellent precursor to diamond. This can be attributed in part to the curved structure of fullerenes, which possess a partial sp bonding configuration, as opposed to graphite, which is planar. MER and Argonne National Laboratories Corporation have investigated fullerenes as the source of carbon to CVD diamond with excellent results. MER has demonstrated excellent diamond coatings using fullerenes as the only source of carbon to deposit diamond at five times the growth rate of methane carbon sources, resulting in a projected 70% cost savings.

Optical Applications of Fullerene

Optical limiting refers to a decrease in transmittance of a material with increased intensity of incident light. The phenomenon of optical limiting has a significant potential for applications in sensor and eye protection from intense sources of light. Based on the optical limiting properties of fullerenes, one can make an optical limiter, which permits all light below an activation threshold to pass and maintains the transmitted light at a constant level below the damage threshold for the sensor or the eye.

4.3 Liquid Crystals: Types and Applications

Liquid Crystals (LC)

Liquid crystals are substances that exhibit long range order in one or two dimensions. They are substances that exhibit a phase of matter that has properties between those of a conventional liquid and those of a solid crystal.

A large number of organic molecules with long chain such as cholesteryl acetate $(CH_3COOC_{27}H_{45})$, cholesteryl benzoate $(C_6H_5COOC_{27}H_{45})$, etc., show LC behavior. A liquid crystal may flow like a ordinary liquid, but have the molecules in the liquid oriented and arranged in a crystal like way or they show anisotropy and double refraction like the crystalline solids.

A liquid crystal is a thermodynamic stable phase characterized by anisotropy of the properties without the existence of a three dimensional lattice, generally lying in the temperature range between isotropic liquid phase and anisotropic solid phase, hence the term mesophase.

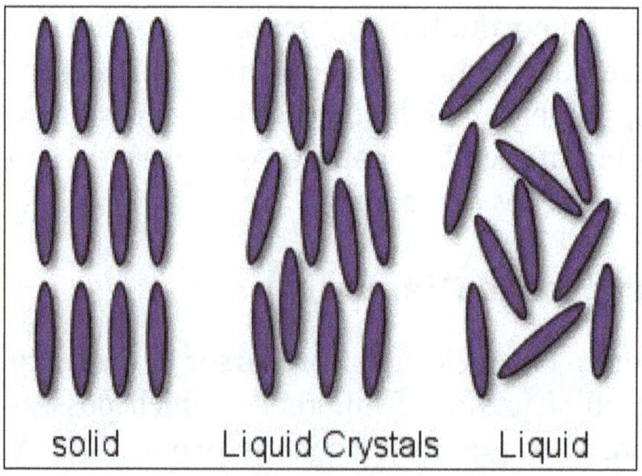

Solid, Liquid crystal and Liquid.

Liquid crystal materials usually have several common characteristics. Among these are rods like molecular structures, rigidness of the long axis and strong dipoles or easily polarizable substituents. The distinguishing characteristic of the liquid crystalline state is the tendency of the molecule to point along a common axis called the director. This is in contrast to molecules in the liquid phase, which have no intrinsic order.

In solid state, molecules are highly ordered and has little translational freedom. The characteristic orientational order of the liquid crystal state is between the traditional solid and the liquid phases and this is the origin of the term mesogenic state, used synonymously with liquid crystal state.

It is sometimes difficult to determine whether a material is in crystal or liquid crystal state. The crystalline materials demonstrates the long range periodic order in three dimensions.

Types

Liquid crystals can be classified into two main categories:

1. Thermotropic Liquid Crystals

It can be classified into two types. They are:

- Enantiotropic liquid crystals.
- Monotropic liquid crystals.

Enantiotropic liquid crystals are those, which can be changed into liquid crystal state by either lowering the temperature of a liquid or raising the temperature of a solid. Monotropic liquid crystals are those, which can only be changed into the liquid crystal state by either increasing the temperature of a solid or decreasing the temperature of a liquid, but not both.

In general, thermotropic mesophases occur because of packing interactions and anisotropic dispersion forces between the molecules.

2. Lyotropic Liquid Crystals

It occurs with the influence of solvents, not by a change in temperature. Lyotropic mesophases occur as a result of solvent induced aggregation of the constituent mesogens into micellar structures.

The lyotropic mesogens are typically amphiphilic, which means that they are composed of both lyophilic and lyophobic parts. This makes them to form into micellar structures in the presence of a solvent, since lyophobic ends will stay together as the lyophilic ends extend outward towards the solution.

As the concentration of the solution is increased and it is cooled, the micelles increase in size and eventually coalesce. This separates newly formed liquid crystalline state from the solvent.

Liquid Crystal Displays

Liquid crystals are widely use in LCDs, which rely on the optical properties of certain liquid crystalline molecules in the presence or absence of electric field. In a typical device, a liquid crystal layer sits between two polarizers that are crossed (i.e.) oriented at 90° to one another.

The liquid crystal is so chosen, that its relaxed phase is a twisted one. This twisted phase reorients light that has passed through the first polarizer, allowing it to be transmitted through the second polarizer and reflected back to the observer. Thus, the device appears clear.

When an electric field is applied to the LC layer, all the mesogens align. In this aligned state, mesogens do not reorient light, so the light polarized at the first polarizer is absorbed at the second polarizer and the entire device appears dark.

In this way, electric fields can be used to make a pixel switch between clear or dark, on command. The colour LCD system uses the same technique, with colour filters used to produce green, blue and red pixels. Similar principles can be used to make other liquid crystal based optical devices.

A liquid crystal display consists of an array of tiny segments that can be manipulated to present information. This basic idea is common to all displays, ranging from a full colour LCD television to simple calculators.

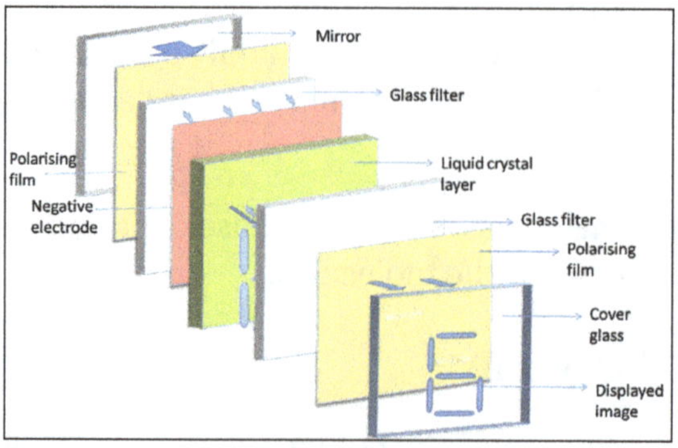

Liquid Crystal Displays.

Importance of Liquid Crystal Displays

The first factor is size. An LCD primarily consists of two glass plates with some liquid crystal material between them. There is no bulky picture tube. This makes LCDs practical for applications where size is important.

In general, LCDs use much less power than their Cathode Ray Tube counterparts. Many LCDs are reflective, which means that they use only ambient light to illuminate the display. Even the displays that do require an external light source consume much less power than CRT devices.

Liquid crystal displays do have drawbacks, shorter lifetime of LC displays limits their use. Only AC drive can give a lifetime more than 10,000 hours whereas, on DC excitation the maximum lifetime that can be obtained is 3000 hours.

Applications of Liquid Crystal

Liquid crystals have a multitude of other uses:

- They are used for non-destructive mechanical testing of materials under stress. This technique is also used for the visualization of RF waves in waveguides.
- They are used in medical applications. For instance, transient pressure transmitted by a walking foot on the ground is measured.
- Low Molar Mass liquid crystals have applications including erasable optical disks, full colour electronic slides for Computer Aided Drawing and light modulators for colour electronic imaging.
- Some of the liquid crystals are used in hydraulic break or clutch system due to their high viscosity values.
- An application of liquid crystals that is now being explored is optical imaging and recording.

4.4 Superconductors: Type-I and Type-II, Properties and Applications

Superconductors

Electrical resistance of metals depends on temperature. Electrical resistance decreases with decreases in temperature and becomes almost zero near the absolute temperature. Materials in this state are said to possess superconductivity. Thus, superconductivity can be defined as a phenomenon in which alloys, chemical compounds and metals become perfect conductors with zero resistivity at temperatures approaching absolute zero.

Superconductors are diamagnetic. The phenomenon was first discovered by Kamerlingh Onnes in 1913 when he found that mercury becomes a superconductor at 4K. The temperature at which a substance starts behaving as superconductor is known as transition temperature, which lies in between the range of 2 to 5K.

Attaining low temperature with liquid helium is highly expensive. The highest temperature at which superconductivity has been observed is 23K for alloys of niobium

(Nb_3Ge), many complex metal oxides have been found to possess superconductivity at fairly high temperatures. Few examples are, $YBa_2Cu_3O_7$ at 90K, $Bi_2Ca_2Sr_2Cu_3O_{10}$ at 105K, $TI_2Ca_2Ba_2Cu_3O_{10}$ at 125K.

Superconductors have many applications in aviation transportation, building magnets, electronics and power transmission.

Classification

Depending upon their behaviour in an external magnetic field, superconductors are divided into two types. They are:

- Type I superconductors.
- Type II superconductors.

Type I Superconductor

In type I superconductor, the magnetic field is completely excluded from the material below the critical magnetic field and the material loses its superconducting property abruptly.

Characteristics

- They exhibit complete Meissner effect.
- They have only one critical magnetic field value.
- Below the critical magnetic field, the material behaves as superconductor and above the critical magnetic field, the material behaves as normal conductor.
- These are called as soft superconductors.

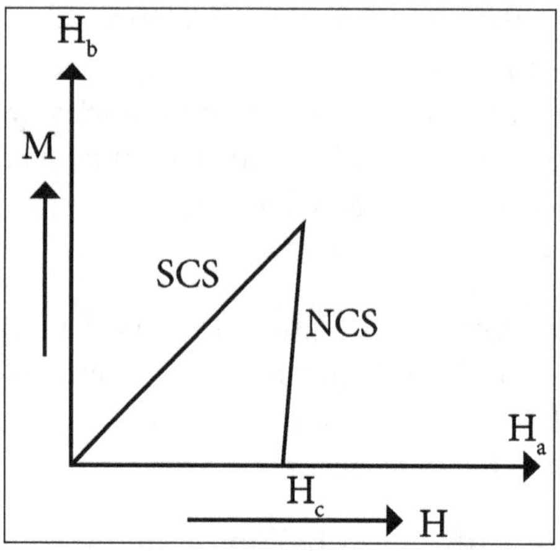

Type I superconductor.

Type II Superconductor

In type II superconductor, the magnetic field is excluded from the material and the material loses its superconducting property gradually rather than abruptly.

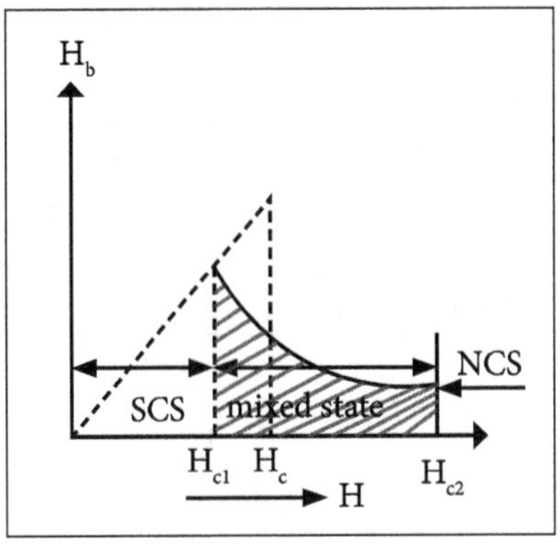

Type II superconductor.

Characteristics

- They do not exhibit a complete Meissner effect.
- They have two critical magnetic field values. Lower critical magnetic field $[H_{C1}]$ and higher critical magnetic field $[H_{C2}]$.
- Below H_{C1}, the material behaves as superconductor and above H_{C1}, the material behaves as normal conductor. The region in between $[H_{C1}]$ and $[H_{C2}]$ is known as vortex region or mixed state.
- These are called as hard superconductors.

Difference between type I and II superconductors

S. No.	Type I (or) soft superconductors	Type II (or) hard superconductors
1.	It exhibits complete Meissner effect.	It does not exhibit a complete Meissner effect.
2.	It is completely diamagnetic.	It is not completely diamagnetic.
3.	Example: Tin, Lead, Mercury, etc.	Example: Nb-Zr, Nb-Ti, Nb-Sn, Va-Ga, etc.
4.	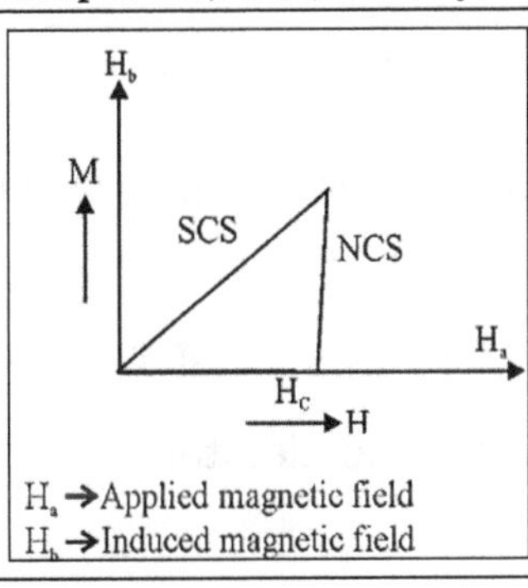	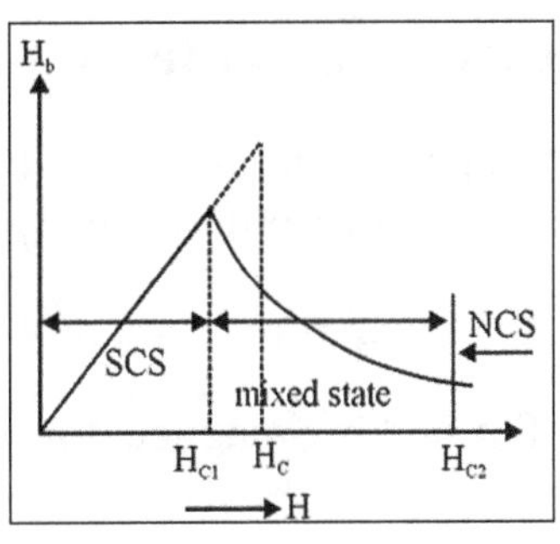

5.	The value of HC is always too low. It is about 0.1 tesla.	The values of H_C is high and it is about 30 tesla.
6.	It has only one critical magnetic field value.	It has two critical magnetic field values. Lower critical magnetic field [HC1] and higher critical magnetic field $[H_{C2}]$.
7.	No mixed state exists.	Mixed state exist.
8.	These are called soft superconductors.	These are called hard superconductors.
9.	It requires low magnetic field to destroy superconductivity.	It requires large magnetic field to destroy superconductivity.
10.	It loses its magnetization abruptly.	It loses its magnetization gradually.

Characteristics of Superconductors

The properties of superconductors are as follows:

- There are small changes in thermal conductivity and volume of materials.
- The electrical resistance can be decreased by lowering the temperature and at transition temperature, the resistance abruptly becomes close to zero.
- There is discontinuous change in specific heat.
- The magnetic flux lines are excluded from the superconductors.
- The transition temperature is the characteristic of a particular isotope of an element.
- Transition temperature of superconductors decreases with increase in atomic mass of isotopes of an element.
- Presence of impurities in superconductors lowers their critical temperature (T_c).
- Below critical temperature, they conduct electricity without change of energy, (i.e.) $\Delta H = 0$. Therefore, current can flow through them continuously.
- All superconductors becomes diamagnetic and are repelled by magnets below critical temperature (T_c) and do not allow magnetic field to pass through it.

Applications of Superconductors

- Electric generators can be made by using superconductors with smaller size, less weight and low energy consumption.
- DC superconducting motors are used in ship propulsion and in large mills.
- For making magnets for high energy particle accelerators.

- In high speed computers, called super computers for signal transmission, high memory storage and for other purposes.
- Used in Magnetic Resonance Imaging for medical diagnosis.
- Superconductors can be used in switching devices.
- They can be used for the detection of brain tumor, defective cells, etc.
- They can also be used for the transmission of power over very long distances.
- They can be used in sensitive electrical instruments for detecting very small direct currents and voltages.
- These are used in NMR (Nuclear Magnetic Resonance) imaging equipments which is used for scanning purposes.
- These are used to design Cryotron, Maglev, Josephson Devices and SQUID.
- Superconducting magnetic fields may be used to launch satellite into orbit directly from the earth without the use of rockets.
- Superconducting solenoids are used in magnetohydrodynamic power generation to maintain the plasma in the body.

4.5 Green Synthesis: Principles, 3 or 4 Methods of Synthesis with Examples and R4M4 Principles

Green synthesis approaches includes mixed valence tollens, polysaccharides, polyoxometalates, biological and irradiation methods which have advantages over conventional methods involving chemical agents associated with environmental toxicity. Selection of ecofriendly non-toxic reducing and stabilizing agents and selection of solvent medium are the most important issues which should be considered in green synthesis of NPs.

Principles of Green Chemistry

1. Atom Economy

Synthetic methods should be designed to maximize the incorporation of all materials used in the process into the final products.

2. Catalysis

Catalytic reagents are superior to stoichiometric reagents.

3. Design for Degradation

Chemical products should be designed so that at the end of their function, they break-down into innocuous degradation products and do not persist in the environment.

4. Design for Energy Efficiency

Energy requirements of chemical processes must be recognized for their environmental and economical.

impacts and should be minimized. If possible, synthetic methods should be carried out at ambient temperature and pressure.

5. Designing Safer Chemicals

Chemical products should be designed to affect their desired function while minimizing their toxicity.

6. Inherently Safer Chemistry for Accident Prevention

Substances and the form of a substance used in a chemical process should be chosen to reduce the changes for chemical accidents, including explosions and fires.

7. Less Hazardous Chemical Syntheses

Wherever practicable, synthetic methods should be designed to use and generate substances that possess little or no toxicity to environment and human health.

8. Prevention

It is better to prevent waste than to clean up or treat after it has been created.

9. Reduce Derivatives

Unnecessary derivatization should be minimized or avoided, because such steps requires additional reagents and can generate waste.

10. Real Time Analysis for Pollution Prevention

Analytical methodologies need to be developed further to allow for real time, in process monitoring and control, prior to the formation of hazardous substances.

11. Safer Solvents and Auxiliaries

The use of auxiliary substances such as separation agents, solvents, etc., should be reduced wherever possible and innocuous when used.

12. Use of Renewable Feedstocks

A feedstock or raw material should be renewable rather than depleting whenever economically and technically practicable.

Of the various principles of green chemistry, the important one is maximizing the atom economy which determines the efficiency of chemical transformation and is calculated as,

$$\% \text{ Atom utilization} = \frac{\text{Molecular weight of desired product} \times 100}{\text{Molecular weight of (desired product + waste product)}}$$

Physical Methods

Laser ablation and evaporation-condensation are the most important physical approaches. Physical synthesis of silver NPs makes use of a tube furnace at atmospheric pressure. A typical tube furnace consumes large power of more than several kilowatts and a pre-heating time of several tens of minutes to reach a stable operating temperature. Silver NPs could be synthesized with the help of a small ceramic heater with a local heating area. The small ceramic heater is used to evaporate source materials. The evaporated vapour can cool at a suitable rapid rate because; the temperature gradient in the vicinity of the heater surface is very steep in comparison with that of a tube furnace.

This makes the formation of small NPs possible in high concentration. The particle generation is very stable because, the temperature of the heater surface does not fluctuate with time. This physical method can be used as a calibration device for nanoparticle measurement and to generate nanoparticles for long term experiments. The results showed that the geometric standard deviation, the geometric mean diameter and the total number concentration of NPs increases with heater surface temperatures. Even at high concentration, spherical NPs without agglomeration were observed, with high heater surface temperatures. The geometric standard deviation and the geometric mean diameter of silver NPs ranges from 1.23-1.88 nm and 6.2-21.5 nm respectively.

Advantages

The advantages of physical synthesis methods are as follows:

- There is no solvent contamination in the prepared thin films.
- Distribution of NPs is uniform.

Disadvantages

- Consumes a great amount of energy while raising the environmental temperature around the source material.
- Requires a lot of time to achieve thermal stability.

- Tube furnace occupies a large space.

Chemical Methods

Chemical Reduction

The most common method for synthesis of silver NPs is chemical reduction by organic and inorganic reducing agents. In general, different reducing agents such as ascorbate, elemental hydrogen, sodium borohydride ($NaBH_4$), sodium citrate, N, N-dimethylformamide (DMF), tollens reagent, polyol process and poly (ethylene glycol)-block co-polymers are used for reduction of silver ions in either aqueous or non-aqueous solutions. These reducing agents reduce Ag^+ and leads to the formation of metallic silver (Ag^o), which is followed by agglomeration into oligomeric clusters. These clusters eventually leads to the formation of metallic colloidal silver particles.

It is important to use protective agents to stabilize dispersive NPs during the preparation of metal nanoparticle and to protect the NPs that can be absorbed on or bind onto nanoparticle surfaces thereby, avoiding their agglomeration. The presence of surfactants such as acids, alcohols, amines and thiols for interactions with particle surfaces can protect particles from agglomeration, sedimentation or losing their surface properties and stabilize particle growth.

Electrochemical Synthetic Method

Electrochemical synthetic method can be used to synthesize silver NPs. It is possible to control particle size either by adjusting the electrolysis parameters or to improve homogeneity of silver NPs by changing the composition of electrolytic solutions. Polyphenylpyrrole coated silver nanospheroids (3-20 nm) can be synthesized by electrochemical reduction at the liquid/liquid interface. This nano-compound can be prepared by transferring the silver metal ion from aqueous phase to the organic phase, where it reacts with pyrrole monomer.

Furthermore, spherical silver NPs with narrow size distributions can be conveniently synthesized in aqueous solution by electrochemical method. Poly N-vinylpyrrolidone was chosen as the stabilizer for silver clusters. Poly N-vinylpyrrolidone protects NPs from agglomeration which significantly reduces silver deposition rate, silver particle formation rate and promotes silver nucleation. Addition of sodium dodecyl benzene sulphonate to the electrolyte improves particle size and particle size distribution of silver NPs.

Irradiation Methods

Silver nanoparticles can be synthesized by using a variety of irradiation methods. Laser irradiation of an aqueous solution of silver salt and surfactant can produce silver NPs with a well-defined shape and size distribution. Laser is being used in photo-sensitization synthetic method for making silver NPs using benzophenone. At short irradiation times, low laser powers produced silver NPs of about 20 nm, whereas, an increase in irradiation power produced NPs of about 5 nm.

Mercury lamps and lasers can be used as light sources for production of silver NPs. In visible light irradiation, photo-sensitized growth of silver NPs using thiophene and silver nanoparticle formation by illumination of $Ag(NH_3)^+$ in ethanol has been done.

Scheme of Synthesis

1. Scheme of synthesis and mechanism by conventional method

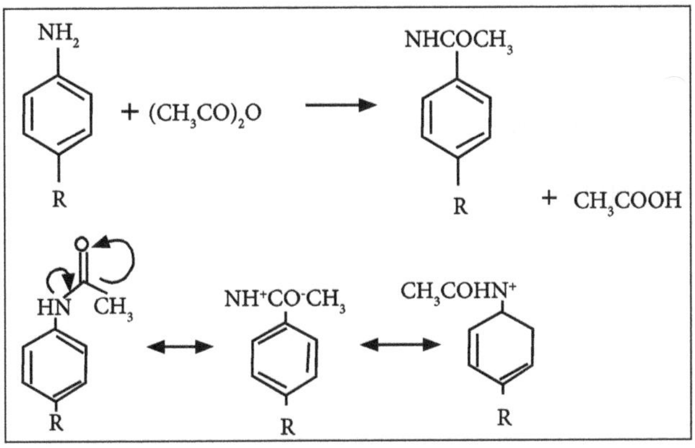

2. Scheme of synthesis and mechanism by green chemistry method

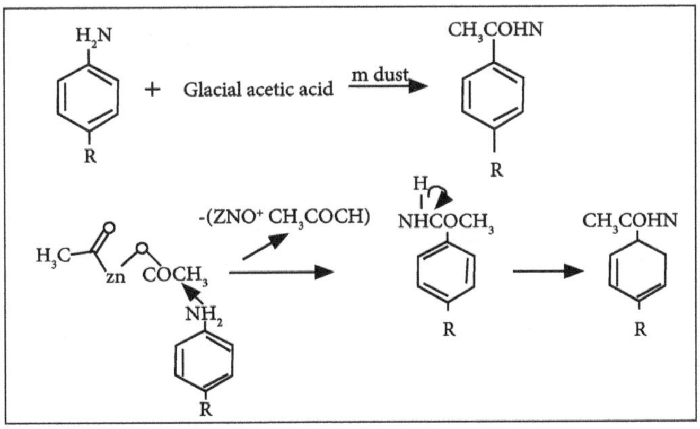

R4M4 Principle

R4M4 means Reduce, Reuse, Recycle, Redesign, Multipurpose, Multidimensional, Multifaceted and Multitracking.

Green Chemistry

Green chemistry is the new branch of chemistry which involves pulling together techniques, tools and technologies. It is helpful to chemists and chemical engineers in the research, development and production of more ecofriendly and efficient products which may also have significant financial benefits.

The development of green chemistry redefines the role of a solvent. An ideal solvent facilitates the mass transfer but does not dissolve. Moreover, a desirable green solvent should be cheap, natural, non-toxic and readily available with additional benefits of aiding the reaction, separation or catalyst recycling.

Applications of Green Chemistry

- As starting materials.
- In agriculture.
- In manufacture of drugs.
- In green chemical products.

5

Solid State Chemistry

5.1 Types of Solids

Solid

A solid is defined as that form of matter which possesses rigidity and hence possesses a definite shape and volume. Free movement of particles in the gaseous state and in the liquid state is the reason for fluidity which is due to weaker intermolecular forces whereas, the particles in the solid state are held together with strong intermolecular forces and do not possess translator motion. They can vibrate only about their fixed position. Therefore, solids are rigid and have definite shape.

Types

Solids are classified into two main types:

- Crystalline solids.
- Amorphous solids.

Crystalline solids

These are substances whose constituent particles possess a regular orderly arrangement. They are considered as true solids. All solid metals, non-metals and compounds come under this category. Few examples are: diamond, NaCl, sucrose, quartz, etc.

Properties

The main properties of these solids are mentioned below:

1. Arrangement of constituents particles

In a crystalline solid, the particles such as atoms, ions and molecules are arranged in definite three dimensions geometric pattern. This is known as long range order. This arrangement repeats periodically over the entire crystal. Due to this arrangement, they have both short and long range orders.

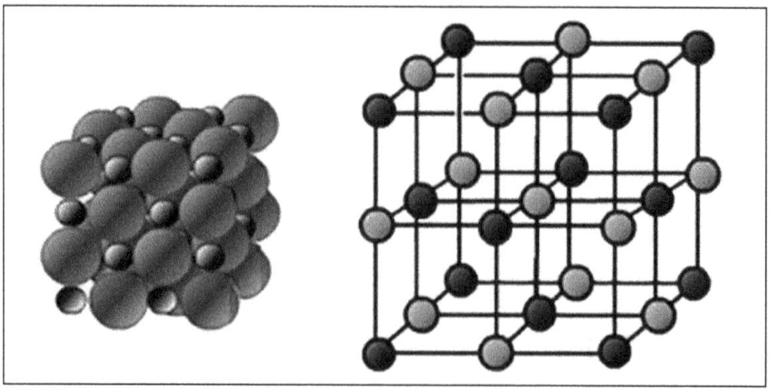

2. Melting Point

The crystalline solids have a sharp melting point, so they possess definite heat of fusion.

3. Anisotropy

In a crystalline solid, the properties like thermal expansion, refractive index, electrical conductance, etc., have different values in different directions. This type of behaviour is known as anisotropy and the substances with this property are called anisotropic.

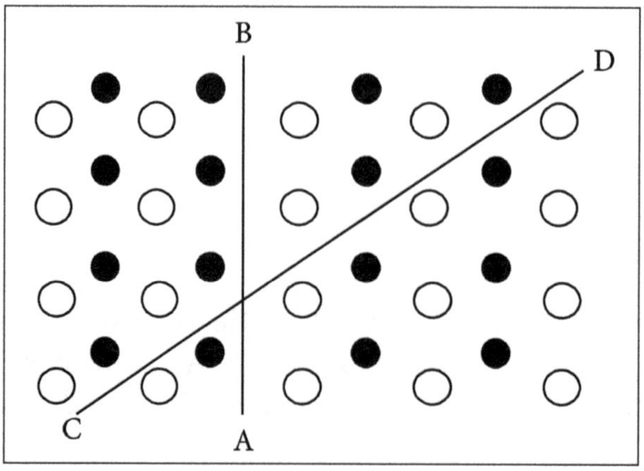

4. Clean cleavage with Knife

A crystalline solid gives a clean surface after cleaving it with a knife rather than an irregular breakage.

Amorphous Solids

These are substances whose constituent particles do not possess a regular orderly arrangement. Glass, pitch, polymers of high molecular mass, rubber, quartz glass, etc., are examples of it.

Properties

The main properties of these solids are described below:

1. Arrangement of Constituents Particles

In an amorphous solid, the particles are arranged in a regular manner up to a small region only. This is known as short range order. So, in these solids, all the particles are not arranged in a regular manner. They possess only short range order and have irregular shape.

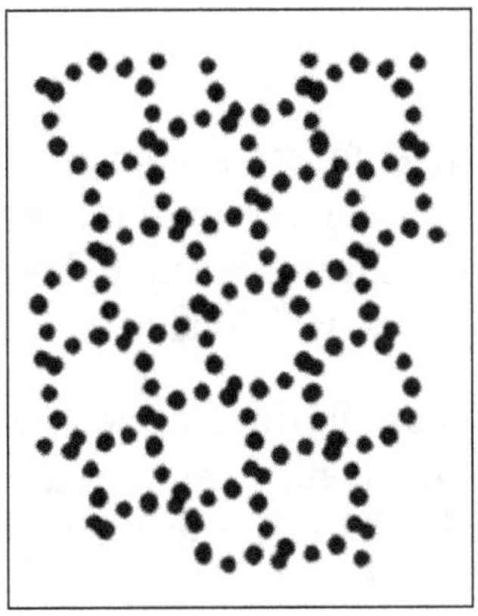

2. Melting Point

The amorphous solid melts gradually over a temperature range, so they do not possess definite heat of fusion.

3. Isotropy

In amorphous solids, the various physical properties like refractive index, thermal expansion, electrical conductivity, etc., are same in all directions just like liquids and gases. This type of behaviour is termed as isotropy and the substances with this property are known as isotropic. So they are isotropic in nature due to random arrangement of particles.

4. No Clean Cleavage with Knife

Like crystalline solids, they do not give a clean surface after cleavage with knife. They undergo an irregular breakage.

5. Super Cooled Liquids

They have a tendency to flow like liquids, although very slowly, so they are called as super cooled liquids. This tendency is illustrated by the fact that, glass panes in the window of an old historical building are found to be thicker at the bottom than at the top.

6. Pseudo Solids

They are not considered as true solids. Due to short range order, small parts of amorphous solids may be crystalline and rest may be non-crystalline. So they are also known as pseudo solids. This is the reason for the milkiness of glass panes in the windows of old buildings because, due to heating at the day and cooling at night, glass acquires some crystalline characters.

5.1.1 Close Packing of Atoms and Ions

Close Packing

Figure shows two possible arrangements for a layer of identical atoms. On squeezing the square layer in figure (a), the spheres would move to the positions in figure (b) so that the layers takes up less space. Layer A is said to be close packed. The hollow spaces between the spheres are of two types, which have been marked with dots and crosses.

To build up a close packed structure in three dimensions, we must add a second layer that is layer B. The spheres of the second layer can sit only in half of the hollows of the first layer and layer B in figure sits directly over the hollows marked with a cross.

When we add a third layer, there are again two possible options as to where it can go. First, it could go directly over layer A, in the unmarked hollows, if we repeat this stacking sequence, we would build up the layers ABABABA... and so on. This is known as Hexagonal Close Packing. In this structure, the hollows marked with a dot are never occupied by spheres, leaving very small channels through the layers.

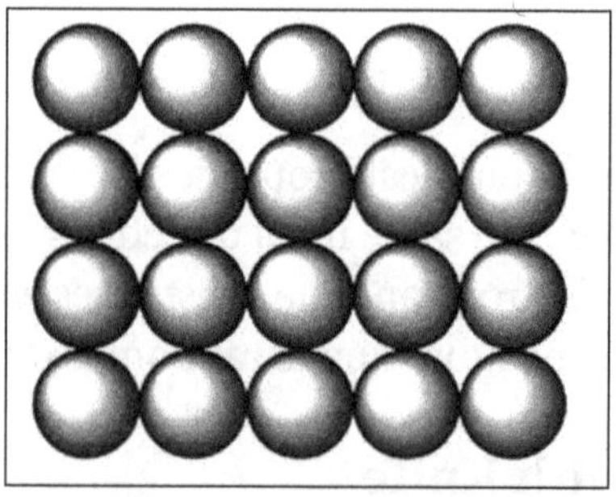

(a)

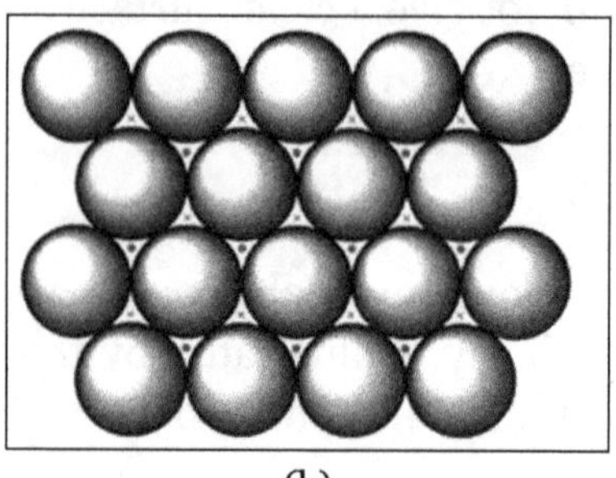

(b)

(a) A square array of spheres (b) A close packed layer of spheres.

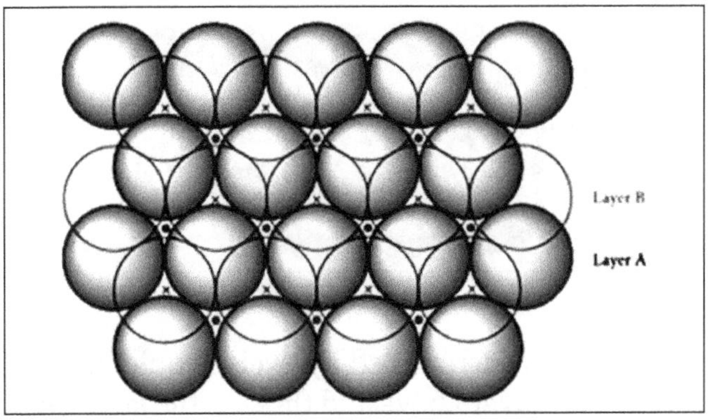

Two layers of close packed spheres.

In the second option, the third layer could be placed over those hollows, as in figure (b), marked with a dot. This third layer, which we now label as layer C, would not be directly over either layer A or B and the stacking sequence when repeated would be ABCAB-CABC... and so on. This is known as Cubic Close Packing. Close packing represents the most efficient use of space, when packing identical spheres 74% of the total volume is occupied by the spheres hence, the packing efficiency is said to be 74%.

(a)

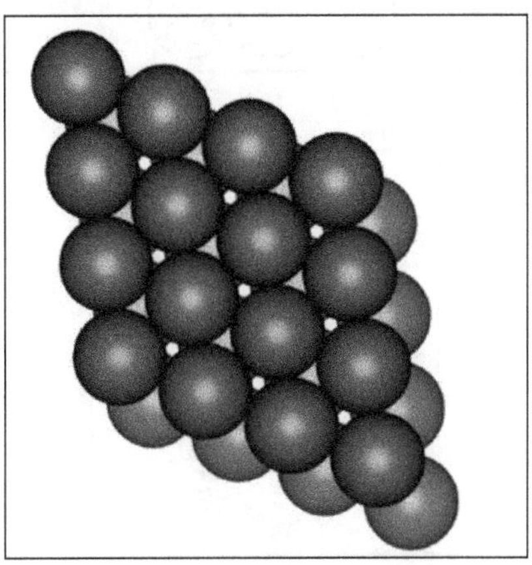

(b)

(a) Three HCP layers showing ABAB. Stacking sequence (b) Three HCP layers showing the narrow channels through the layers.

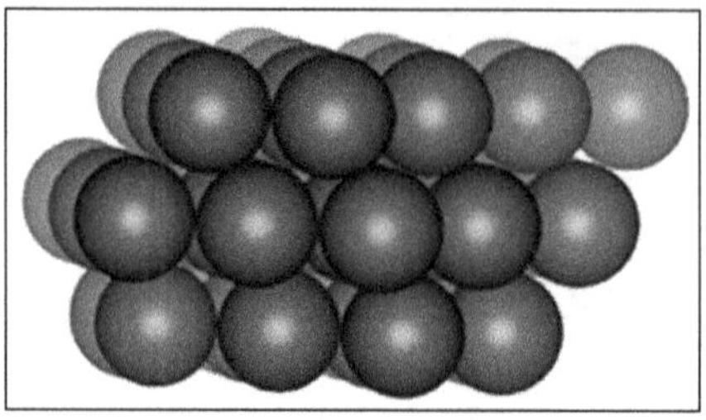

Three CCP layers.

Each sphere in the structure is surrounded by twelve equidistant neighbours (i.e.) six in the same layer, three in the layer above and three in the layer below. Thus, the co-ordination number of an atom in a close packed structure is 12. Another important feature of close packed structure is the number and shape of the small amounts of space trapped in between the spheres. There are two different types of spaces within a close packed structure. The first is called an octahedral hole and the second is known as tetrahedral hole. Figure (a) shows two close packed layers, but with the octahedral holes shaded. Each of these holes is surrounded by six spheres, three in layer A and three in layer B, the centers of these spheres lies at the corners of an octahedron, hence the name. If there are n spheres in the array, then there are n octahedral holes.

Similarly, figure (a) shows two close packed layers, now with the second type of space, tetrahedral holes, shaded. Each of these holes is surrounded by four spheres with centers at the vertices of a tetrahedron. If there are n spheres in the array, then there are 2n tetrahedral holes.

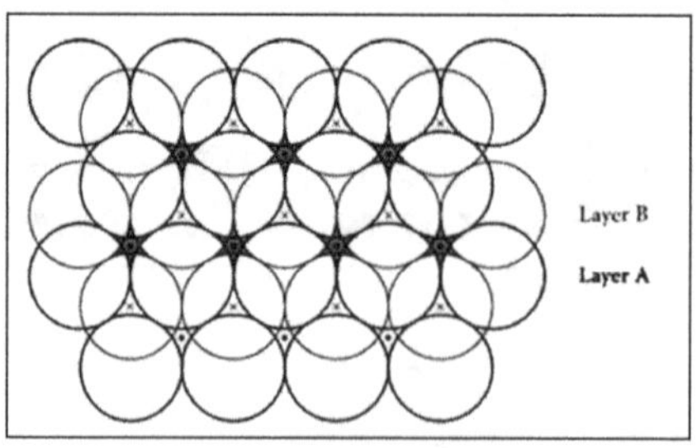

(a)

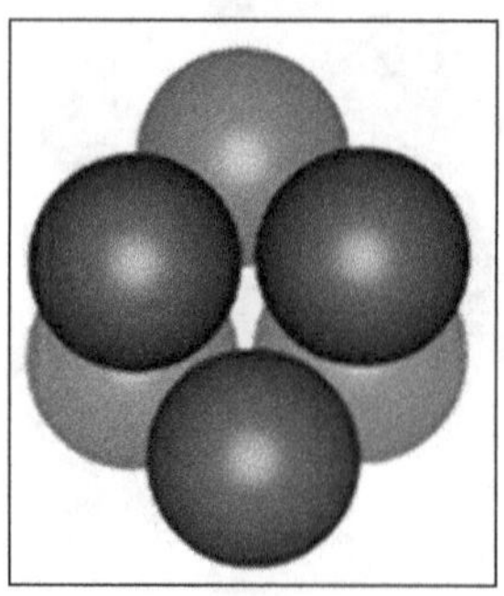

(b)

(a) Two layers of close packed spheres with the enclosed octahedral holes shaded
(b) A computer representation of an octahedral hole.

The octahedral holes in a close packed structure are much bigger than the tetrahedral holes they are surrounded by six atoms rather than four. It is a matter of simple geometry to calculate the number of atoms in a close packed array of spheres of radius r, the radius of a sphere that will just fit in an octahedral hole is 0.414r. For a tetrahedral hole, the radius is 0.225r.

There are, innumerable stacking sequences possible when repeating close packed layers. However, hexagonal close packed structures and cubic close packed are those of maximum simplicity and are commonly used in the crystal structures of the metallic elements and of the noble gases. Only two other stacking sequences are found in perfect crystals of the elements, an ABAC repeat in La, Pr, Nd and Am and a nine layer repeat ABACACBCB in Sm.

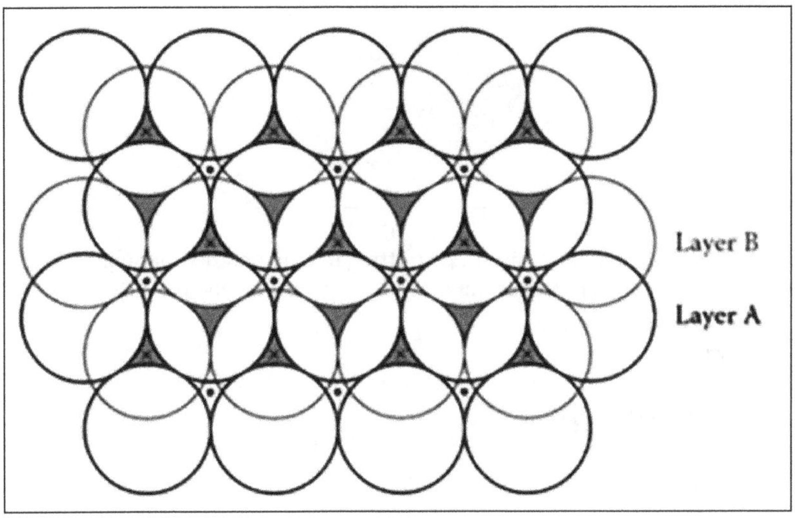

(a)

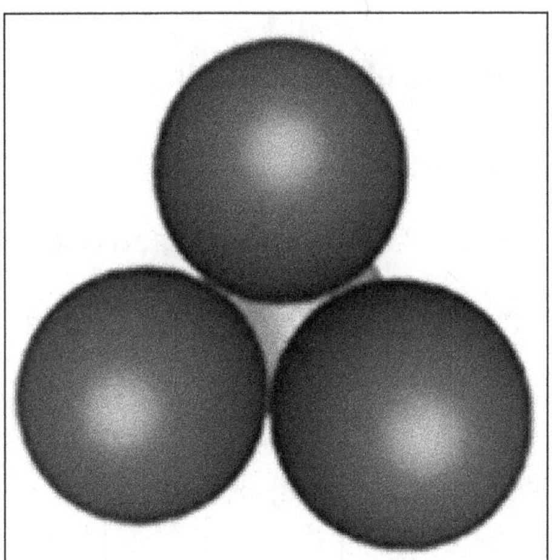

(b)

(a) Two layers of close packed spheres with the tetrahedral holes shaded.
(b) A computer representation of a tetrahedral hole.

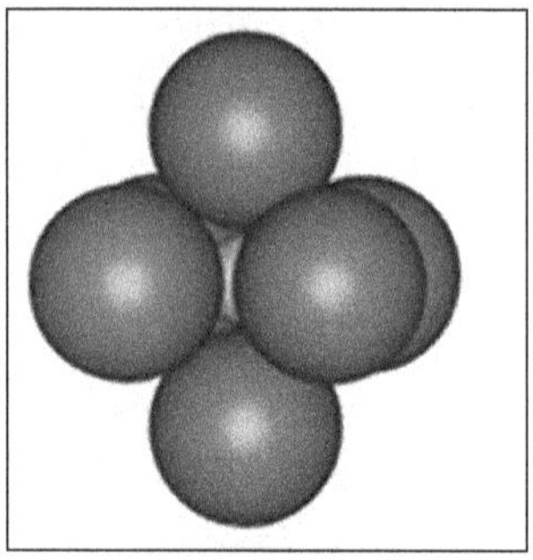

(a)

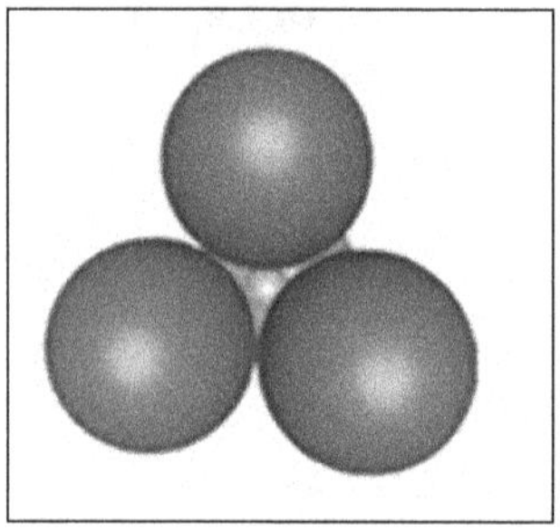

(b)

(a) A sphere radius 0.414r fitting into an octahedral hole
(b) A sphere radius 0.225r fitting into an tetrahedral hole.

5.1.2 BCC, FCC

Types of Cubic Lattices

It has three possible types:

1. Simple Cubic Lattice

It has lattice points at all 8 corners of the unit cell.

2. Body Centered Cubic Lattice

It has lattice points at all 8 corners of the unit cell and one lattice point at the center of the body as shown below:

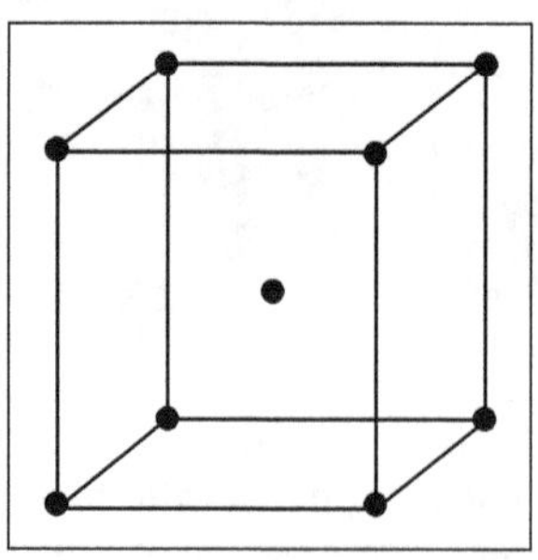

BCC

3. Face centered cubic lattice

It has lattice points at all 8 corners of the unit cell and one lattice point at the center of the 6 faces of the cube.

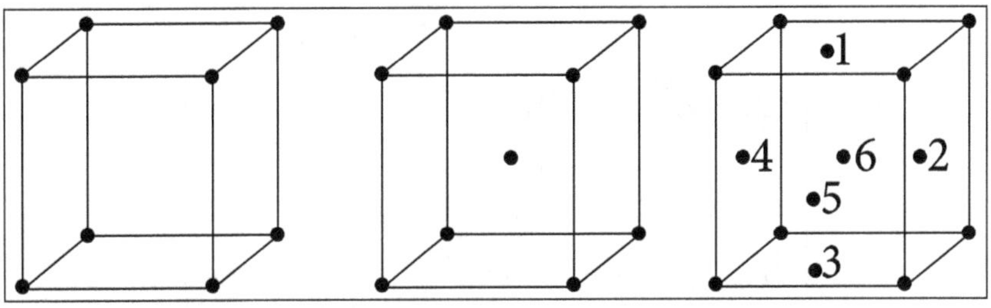

Simple BCC FCC.

System	Bravais lattice	Unit cell characteristics	Characteristic symmetry elements	Examples
Cubic	Simple Body centered Face centered	$a = b = c$ $\alpha = \beta = \gamma = ^\circ 90^\circ$	Four 3 fold rotation axes (along cube diagonal)	NaCl, CaF_2, $NaClO_3$
Tetragonal	Simple Body centered	$a = b \neq c$ $\alpha = \beta = \gamma = 90^\circ$	One 4 fold rotation axis	$NiSO_4$, SnO_2, indium, tin
Orthorhombic	Simple Base centered Body centered Face centered	$a \neq b \neq c$ $\alpha = \beta = \gamma = 90^\circ$	Three mutually orthogonal 2 fold rotation axes	KNO_3,$BASO_4$, $MgSO_4$, sulphur, topaz
Monoclinic	Simple Base centered	$a \neq b \neq c$ $\alpha = \beta = 90^\circ \neq \gamma$	One 2 fold rotation axis	Na_2SO_4,$FeSO_4$, NO_2SO_3
Triclinic	Simple	$a \neq b \neq c$ $\alpha \neq \beta \neq \gamma \neq 90^\circ$	None	$CuSO_4$,$K_2Cr_2O_7$
Trigonal (Rhombohedral)	Simple	$a = b = c$ $\alpha = \beta = \gamma \neq 90^\circ$	One fold rotation axis	$CaSO_4$,Bi, Sb, calcite
Hexagonal	Simple	$a = b \neq c$ $\alpha = \beta = 90^\circ$ $\gamma = 120^\circ$	One fold rotation axis	Tourmaline, quartz

Packing factor

Packing factor is defined as the ratio of volume of atoms in the unit cell to the volume of unit cell.

BCC Structure

The unit cell of BCC is shown in figure.

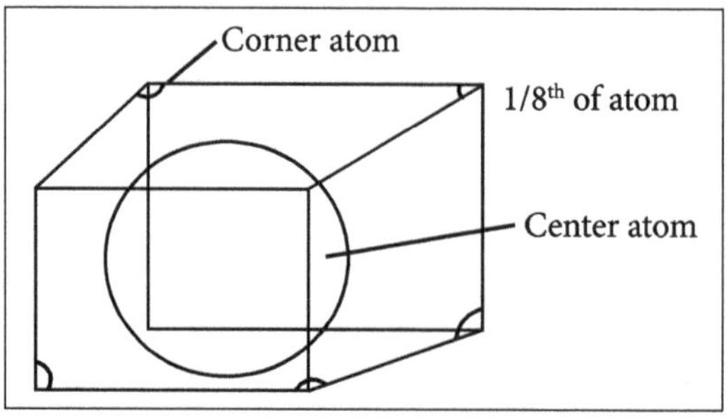

BCC unit cell.

Each corner atom is shared by 8 adjoining or surrounding unit cells.

Share of one unit cell = 1/8 of each corner atom. There are '8' corner atoms.

$$\frac{1}{8} \times 8 = 1 \text{ atom}$$

There is one atom at the body center of every unit cell.

∴ Total number of atom in one unit cell = 1 + 1 = 2 atoms.

Co-ordination Number

In the unit cell of BCC structure, there is one atom at the body center of the unit cell. Further there are '8' atoms at the 8 corners of the unit cell as shown. The corner atoms do not touch each other. But all the eight corner atoms of the unit cell touch the body centered atom along the body diagonal. Thus, for body center atom 'X', there are 8 nearest neighbours.

Hence, the co-ordination number of body centered cubic structure is 8.

Packing Factor

Number of atoms per unit cell = 2

Volume of 2 atoms, $V = 2 \times \frac{4}{3}\pi r^3$

Side of the unit cell, $a = \frac{4r}{\sqrt{3}}$

Volume of the unit cell, $v = a^3$

Packing factor $= \frac{v}{V}$

$$P.F. = \frac{2 \times \frac{4}{3}\pi r^3}{a^3}$$

Substituting for a, we get,

$$P.F = \frac{2 \times \frac{4}{3}\pi r^3}{\left(\frac{4r}{\sqrt{3}}\right)^3} = \frac{\frac{8}{3}\pi r^3}{\frac{4^3 \times r^3}{(\sqrt{3})}} = \frac{\frac{8}{3}\pi r^3}{\frac{64\ r^3}{\sqrt{3}\times\sqrt{3}\times\sqrt{3}}} = \frac{\frac{8}{3}\pi r^3}{\frac{64r^3}{3\sqrt{3}}}$$

$$P.F = \frac{8}{3}\pi r^3 \times \frac{\sqrt{3}\times 3}{64\ r^3}\ ;\ P\cdot F = \frac{\sqrt{3}\,\pi}{8}$$

$$P.F = \frac{\sqrt{3}\times 3.14}{8} = 0.68 = P\cdot F = 68\%$$

FCC Structure

Number of Atoms Per Unit Cell

A unit cell of face centered cubic structure is shown. There are 8 corner atoms, one at each of its 8 corners. Each corner atom is shared by 8 surrounding unit cells.

Hence, the share of each unit is 1/8 of each corner atom.

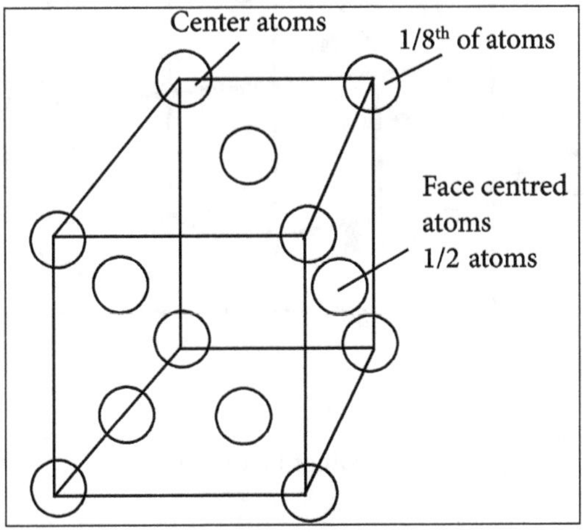

FCC structure.

Number of atoms in the unit cell from the contribution of corner atoms $= \frac{1}{8} \times 8 = 1$ atom.

In addition, there are 6 atoms at the face centers of the cube. Each face centered atom is shared by 2 surrounding unit cells.

Hence, the share of each unit cell = 1/2 of each face centered atom.

∴ Number of atoms, from the contribution of face centered atom = $1/2 \times 6 = 3$ atoms

∴ Total number of atoms in unit cell = $1 + 3 = 4$.

Co-ordination Number

In FCC structure, there are 8 corner atoms and 6 face centered atoms, one at the center of each face of the unit cell.

To determine the co-ordination number, let us consider a corner atom (x) of the unit cell as shown in the below figure. There are three mutually perpendicular planes with a common point of intersection on the atom X. In plane I, it has 4 face centered atoms (1, 2, 3, 4) as nearest neighbours. In plane II, it has 4 more face centered atoms (5, 6, 7, 8) as nearest neighbours for the corner atom X. Similarly plane III has 4 more face centered atom (9, 10, 11, 12) as nearest neighbours to the corner atom X.

The total number of nearest atoms to any corner atom is $4 + 4 + 4 = 12$. Thus, the co-ordination number is 12.

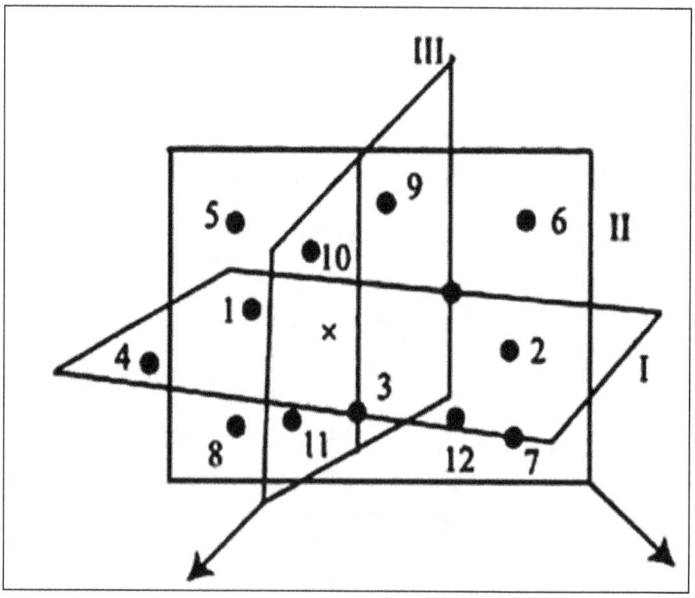

Co-ordination of FCC.

Packing Factor

Number of atoms per unit cell = 4

Volume of 4 atoms $v = 4 \times \frac{4}{3}\pi r^3$

Side of the unit cell, $a = \frac{4r}{\sqrt{2}}$

$$\left[\because \text{atomic radius } r = \frac{\sqrt{2}\,a}{4}\right]$$

Volume of the unit cell,

$$V = a^3 = \left[\frac{4r}{\sqrt{2}}\right]^3$$

$$\text{Packing factor} = \frac{v}{V}$$

Substituting v and V, we have,

$$P.F = \frac{\left(4 \times \frac{4}{3}\pi r^3\right)}{\left(4r/\sqrt{2}\right)^3} = \frac{\frac{16}{3}\pi r^3}{\frac{4^3 r^3}{\left(\sqrt{2}\right)^3}}$$

$$P.F = \frac{\frac{16}{3}\pi r^3}{\frac{64r^3}{\sqrt{2}\sqrt{2}\sqrt{2}}}$$

$$P.F = \frac{\frac{16}{3}\pi r^3}{\frac{64r^3}{2\sqrt{2}}} = \frac{16}{3}\pi r^3 \times \frac{2\sqrt{2}}{64r^3}$$

$$P.F = \frac{\pi\sqrt{2}}{6}\ ;\ P.F = \frac{3.14 \times \sqrt{2}}{6} = 0.74$$

$$P.F = 74\%$$

Problems

1. Consider that the material zinc has HCP structure. If the radius of the atom is 1/4th

of the diagonal of hexagon then, let us calculate the height of the unit cell in terms of atomic radius.

Solution:

Given:

Zinc has HCP structure.

Radius of the atom = 1/4th of the diagonal of hexagon.

Formula to be used:

$r = a/2$

Radius of atom $r = \frac{1}{4}$

$$r = \frac{a}{2} = \frac{1}{4} \Rightarrow a = 2r$$

$$a = 2 \times \frac{1}{4} = \frac{1}{2}$$

w.k.t $\frac{c}{a} = 1.633$

$$c = 1.633\ a$$

2. The density of BCC iron is 7860 kg/m³ and its atomic weight is 58.25. Let us calculate its atomic radius.

Solution:

Given:

Density of BCC iron = 7860 kg/m^3

Atomic weight = 58.25

Formula to be used:

$$a^3 = \frac{nA}{NP}$$

$$r = \frac{a\sqrt{3}}{4}$$

For BCC, n = Z

$$a^3 = \frac{nA}{NP} = \frac{2 \times 58.25}{6.023 \times 10^{26} \times 7860}$$

$$a = 2.87 \times 10^{-10}\ m$$

Atomic radius,

$$r = \frac{a\sqrt{3}}{4}$$

$$r = \frac{2.87 \times 10^{-10} \times \sqrt{3}}{4}$$

$$r = 1.24 \times 10^{-10}\ m$$

$$r = 1.24\ Å$$

5.1.3 Structures of Rock Salt Cesium Chloride: Spinel-Normal and Inverse Spinels

Rock salt structure

- They are MX type compounds based on NaCl or rock salt structure.
- Anions (X) forms the cation sub-lattice with FCC structure.
- Cations (M) fill the octahedral sites.
- There are one octahedral site per anion hence, 100% occupancy of sites according to the stoichiometry.
- Radius ratio, r_c / r_a is in the range of 0.414 - 0.732 with some exceptions.
- Examples of ceramic materials with such structures are NaCl, MgO, NiO, FeO etc.

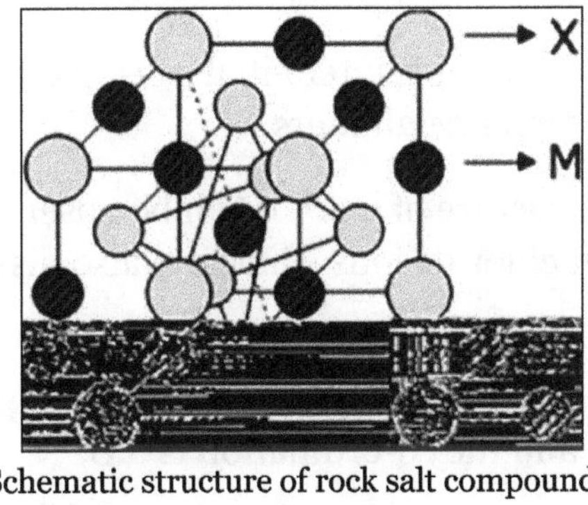

Schematic structure of rock salt compounds.

Cesium Chloride

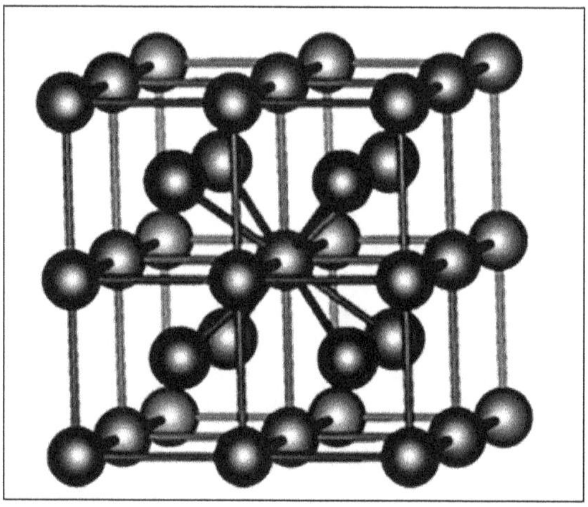

Structure of CsCl.

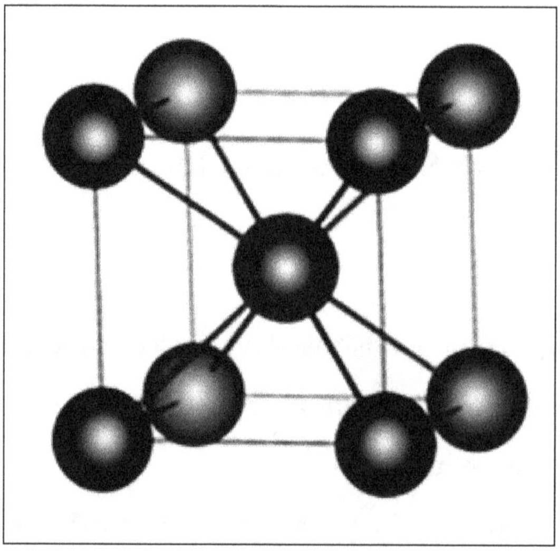

Unit cell representation of CsCl.

Structure

- The cesium ions form the simple cubic arrangement.
- Chloride ions occupy the cubic interstitial sites, (i.e.) each chloride ion has eight cesium ions as its nearest neighbours.
- If we consider the structure of CsCl, it can be observed that each cesium ion is surrounded by eight chloride ions which are also disposed towards the corners of the cube.
- It may be concluded that both type of ions are in equivalent positions and the stoichiometry is 1:1 and the co-ordination is 8:8.

Spinel - Normal and Inverse Spinels

Spinel

Spinel is the name given to the mineral $MgAl_2O_4$. It has a common structural arrangement shared by many oxides of the transition metals with formula AB_2O_4.

In the normal pattern, the oxygen forms the cubic close packed array and the Mg(II) and Al(III) sit in tetrahedral and octahedral sites in the lattice, giving the unit cell with 8 Mgs, 16 Als and 32 Os.

Inverse Spinel

An inverse spinel is an alternative arrangement, where the divalent ions swap with half of the trivalent ions so that the M(II) now occupy octahedral sites (i.e.) $B(AB)O_4$.

The spinels have the general formula AB_2X_4.

Where,

A^{II} = A divalent cation like Mg, Cr, Mn, Fe, Co, Ni, Cu, Zn, Cd, Sn

B^{III} = A trivalent cation like Al, Ga, In, Ti, V, Cr, Mn, Fe, Co, Ni

X = O, S, Se etc.

Structure of Normal Spinels (AB_2O_4)

The divalent A^{II} ion occupies the tetrahedral voids whereas, the trivalent B^{III} ions occupies the octahedral voids in the close packed arrangement of oxide ions. A normal spinel can be represented as, $(A^{II})^{tet}(B^{III})_2^{oct}O_4$.

Example of this are, $MgAl_2O_4$ known as spinel, Mn_3O_4, $ZnFe_2O_4$, $FeCr_2O_4$ etc.

Structure of Inverse Spinels ($B(AB)O_4$)

The A^{II} ions occupies the octahedral voids whereas, half of the B^{III} ions occupies the tetrahedral voids. It can be represented as, $(B^{III})^{tet}(A^{II}B^{III})^{oct}O_4$.

Example of this are, Fe_3O_4, $CoFe_2O_4$, $NiFe_2O_4$ etc.

The above inverse spinels can also be written as,

$$Fe_3O_4 = Fe^{III}(Fe^{II}Fe^{III})O_4$$

$$CoFe_2O_4 = Fe^{III}(Co^{II}Fe^{III})O_4$$

$$NiFe_2O_4 = Fe^{III}(Ni^{II}Fe^{III})O_4$$

The number of octahedral sites occupied may be ordered or random. The random occupation leads to defected spinels.

Example, $NiAl_2O_4$ forwhichtheformulacanbewrittenas $(Al_{0.75}Ni_{0.25})^{tet}[Ni_{0.75}Al_{1.25}]^{octa}O_4$

5.2 Non-Elemental Semiconducting Materials: Stoichiometric, Controlled Valency and Chalcogen Photo/Semiconductors

Non-Elemental Semiconductors

A host of non-elemental materials too exhibit semiconductivity above 0°K. These materials are known as non-elemental or compound semiconductors.

The compounds or intermetallics like GaAs, InSb, PbTe, GaP, etc. and a host of oxide materials such as ZnO, NiO, FeO, V_2O_5 etc., are semiconductors. These compounds are examples of non-elemental semiconductors.

The non-elemental or compound semiconductors are further classified as:

- Stoichiometric semiconductors.
- Non-stoichiometric semiconductors.

1. Stoichiometric Semiconductors

Many stoichiometric intermetallic compounds such as GaAs, GaP, InSb, CdSe, etc., are semiconductors. These intermetallic compounds are synthesized by combining the elements of groups III and V or groups II and VI in 1:1 ratio, they are referred to as stoichiometric semiconductors.

The crystal lattice structures of these stoichiometric intermetallic compounds resembles the crystal structure lattice of pure Si and Ge. Like silicon and germanium, different elements of these compounds also form 4 equivalent covalent bonds with 4 neighbouring elements and crystallize in diamond cubic structure.

The mechanism of electrical conductivity is similar to intrinsic semiconductivity of pure elemental silicon or germanium crystal but these intermetallics have different forbidden energy, E_g values, depending on the nature of constituents.

The intermetallic compounds can also be doped to get n-type or p-type extrinsic semiconductors. These stoichiometric semiconductors exhibit a wide range of intrinsic type semi-conduction and also a wide range of E_g values ranging from 0.5 eV to 2.3 eV. These compounds find extensive application in electronics because of their wide range of E_g values.

Controlled Valency

Thermistors used in power meters consist of a polycrystalline sintered ceramic made from a mixture of oxides chosen from the elements manganese, nickel, cobalt, copper and iron and formed into a small bead. Early thermistor materials suffered from poor reproducibility because, their chemical compositions were sensitive to atmospheric and temperature variations during sintering and subsequent processing, but over the years stability has improved.

In general there are two important conduction mechanisms involved in oxide semiconductors. The first is departure from stoichiometry. For example, conduction in CuO_2 is due to an excess of oxygen, leading to p-type conductivity. The second mechanism is a hopping process which can be controlled by varying the proportions of the constituent oxides. Semiconductors relying on this principle are termed controlled valency semiconductors. Mn_3O_4 is normally an insulator in its pure form, but conduction occurs if nickel is introduced. The hopping mechanism is different from that which applies for semiconductors such as silicon. Instead, electrons hop from one ionic site to the next. When nickel is added to Mn_3O_4 some of the Mn^{3+} ions are induced to change to Mn^{4+} and electrons are then able to hop between Mn^{3+} and Mn^{4+} ions, the Mn^{3+} ion becoming an Mn^{4+} ion and vice versa.

The nickel itself does not take part in conduction. The general conditions for hopping to occur are, the ions between which the electrons jumps must be of the same element must differ in valency by one unit, must occupy crystallographically equivalent sites and must not be too far. The construction of a thermistor bead is shown in below figure. The method of fabrication is as follows. The constituent oxides are first mixed and ground to a homogeneous fine powder, which is then mixed with an organic binder such as polyvinyl acetate to form slurry.

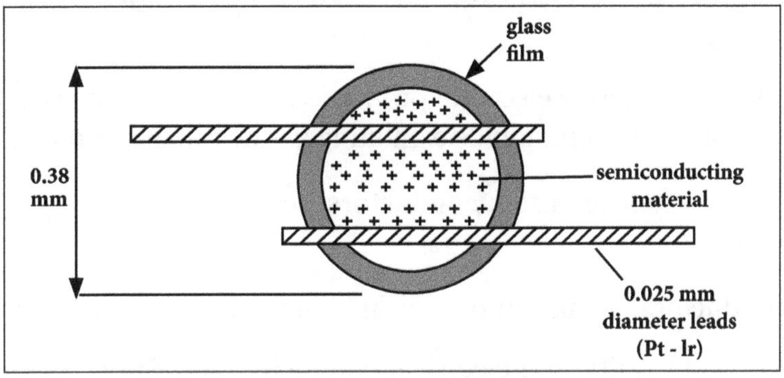

Thermistor bead of the type used in power meters.

A small amount of slurry is applied so as to form a bead of diameter 50-70 μm which bridges two platinum wires of diameter 25 μm. The bead is air dried, sintered in a furnace at 1100-1200°C so that the oxides combine chemically and covered by a protective glass film. The two unwanted lengths of platinum wire are cut off and tungsten tipped nickel lead wires are attached to the two remaining ends. The final stage consists of encapsulation in glass. Ageing improves the stability of the resistance characteristic.

The relationship between the resistance R of the thermistor and the absorbed power P can be represented approximately by the equation,

$$R = J \exp\left(\frac{B}{K+CP}\right) \qquad ...(1)$$

Where J, B, K and C are constants.

If voltage is plotted against current there is usually a region of negative slope resistance over which an increase in current produces a decrease in voltage. If the thermistor is biased in this region, care must be taken to ensure that the circuit is stable. For power measurement the operating resistance normally lies in the range of 100-300 Ω, usually 100 or 200 Ω. The operating temperature is normally slightly higher than 100°C and the maximum power which can be measured is normally 10 mW.

The large temperature co-efficient of resistance of thermistors, which lies in the range of 10-50 Ω/mW, enables the spread in the characteristics of different thermistors to be compensated by adjusting the dc bias current, which happens automatically if a self-balancing bridge is used. Thermistor power meters have reasonably good ability to withstand moderate power overloads, because as the temperature approaches the burn out value the resistance drops sharply to a value typically in the region of 5 Ω, with the result that most of the incident power is reflected.

Chalcogen Photo/Semiconductors

Chalcogen Semiconductors

- The elements of VI group, sulphur, selenium and tellurium are known as chalcogens.
- They have characteristic electrical properties. However, in combination with other elements, they form compounds exhibiting a wide range of interesting properties.
- Among these chalcogen semiconductors, selenium or As_2Se_3 is a good photoconductor.
- Photoconductors are also known as light sensitive semiconductors.
- In photoconductivity, the electrical conductivity of Si or As_2Se_3 is increased

by the impact of a source of light on the sample. The source of energy from an optical beam causes increased charge carriers to be excited across the energy gap, E_g of the photoconductor, as a result of which the conductivity enhances.

Photocopying Process

It is observed that when selenium is exposed to a source of light, its conductivity increases due to the release of electrons. Hence, this property of photosensitivity of selenium is adopted in photocopying process. In a photocopying unit, the surface of a revolving drum acts as a screen, which is smeared with a thin coating of finely powdered selenium.

- In the first process, a corona discharge is induced on the screen of the drum by holding a high potential wire parallel to the rotating drum.
- The document to be photocopied is exposed to a source of light. As the light falls on the screen of the drum, electron-hole pair is formed on the screen for bright areas of the document, while the dark areas of the documents do not affect the screen.
- The drum is discharged under a potential and application of more light, so that the electron moves towards the screen and the holes in the opposite direction.
- Since, light does not pass through the dark areas of the document, a pattern of positive charges (i.e.) an electrical image of the document is formed on the screen. As the drum rotates, carbon ink which has negative polarity is captured by the electrical image on the screen of the drum.
- In the last process, the induced discharge results in the transfer of image with ink from the screen to the plain paper.

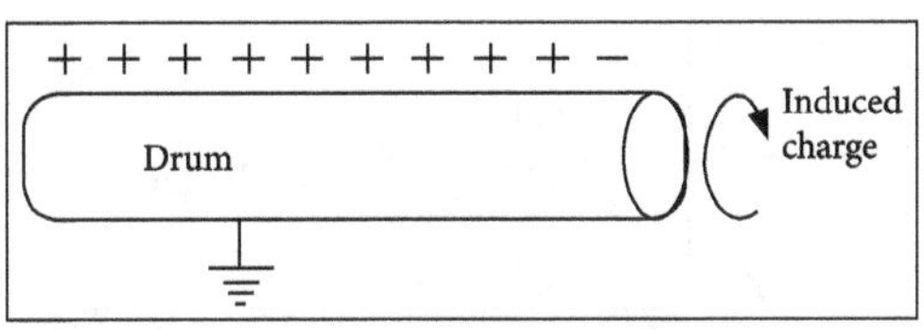

(a)

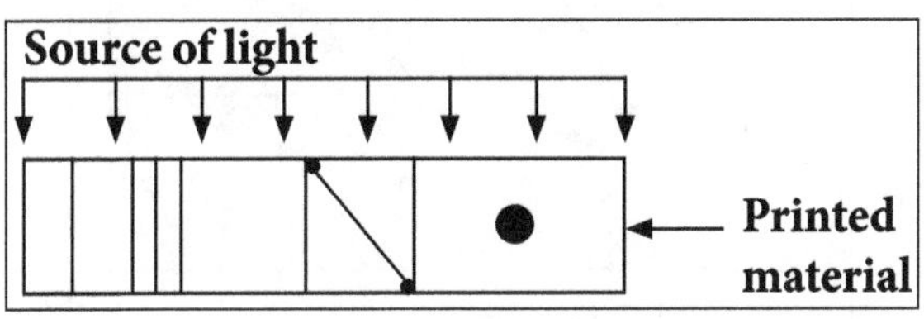

(b)

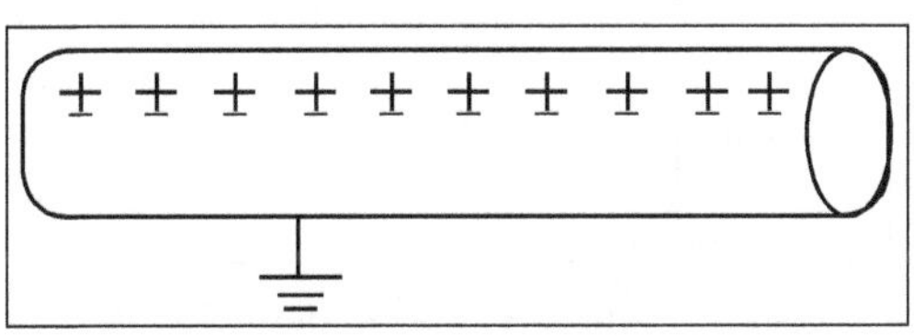

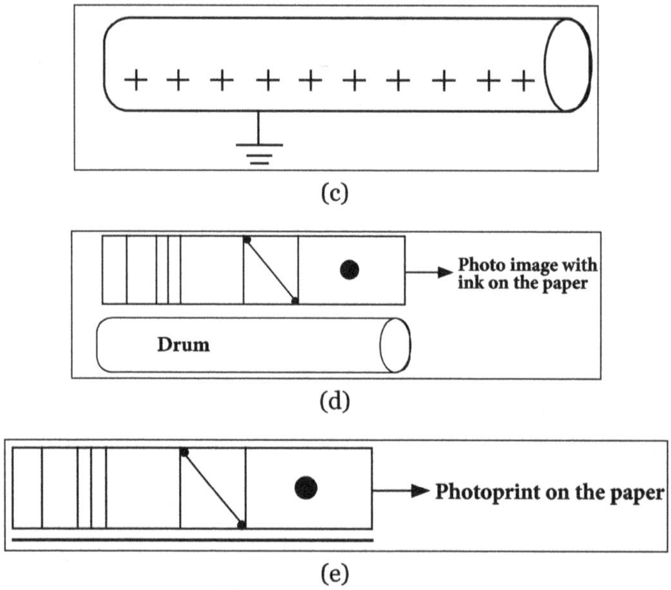

Photocopying process.

5.3 Preparation of Semiconductors and Semiconductor Devices

Basic Concepts of Semiconductor Devices

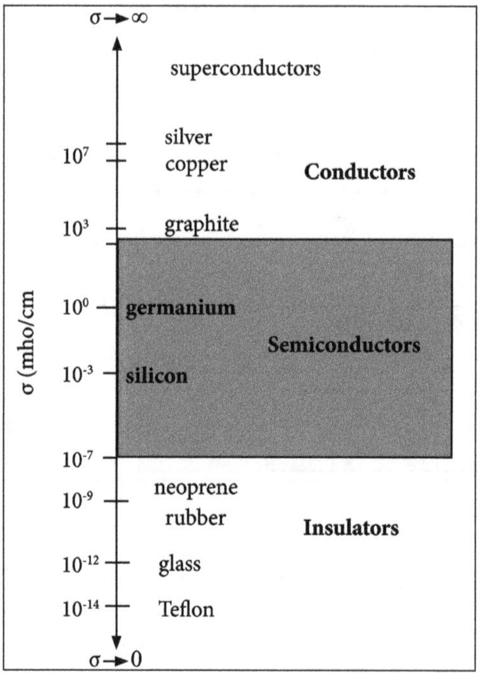

Semiconductors and its conductivity.

Materials that permit flow of electrons are called conductors such as gold, silver, copper, etc. Materials that block flow of electrons are known as insulators. Few examples are rubber, glass, Teflon, mica, etc.

Materials whose conductivity falls between those of conductors and insulators are termed as semiconductors. Semiconductors are part time conductors whose conductivity can be controlled.

Preparation of Semiconductors

Silicon is generally used to build semiconductor devices. Si is the main ingredient of sand and it can be estimated that the cubic mile of seawater contains 15,000 tons of Si. Si is spun and it is grown into a crystalline structure and cut into wafers to manufacture electronic devices.

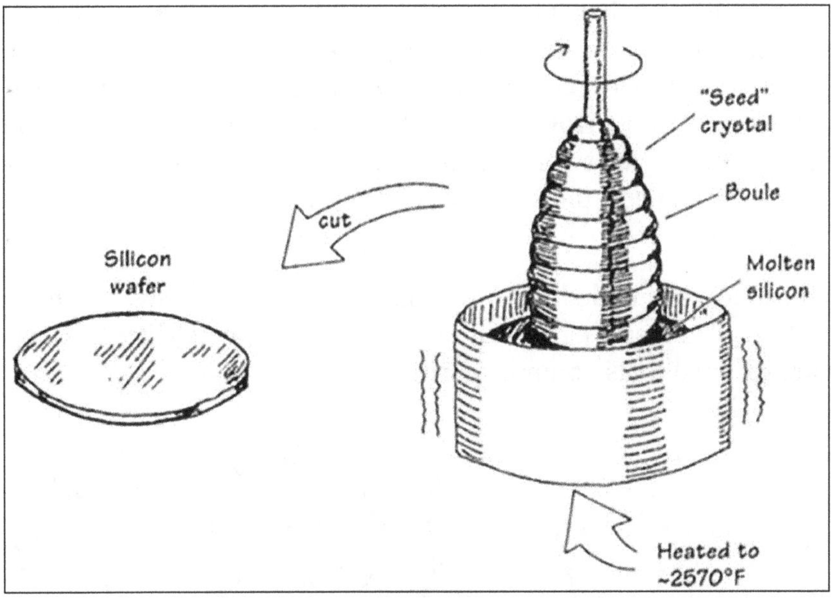

Silicon

Atoms in a pure silicon wafer contains four electrons in the outermost orbit known as valence electrons. Germanium is another semiconductor with four valence electrons. In the crystalline lattice structure of silicon, the valence electrons of every Si atoms are locked up in covalent bonds with the valence electrons of four neighbouring Si atoms.

In the pure form, Si wafer does not contain any free charge carriers. An applied voltage across pure silicon wafer does not yield any electron flow through the wafer. A pure Si wafer act as an insulator.

In order to make useful semiconductor devices, materials such as phosphorus (P) and boron (B) can be added to Si to change Si's conductivity.

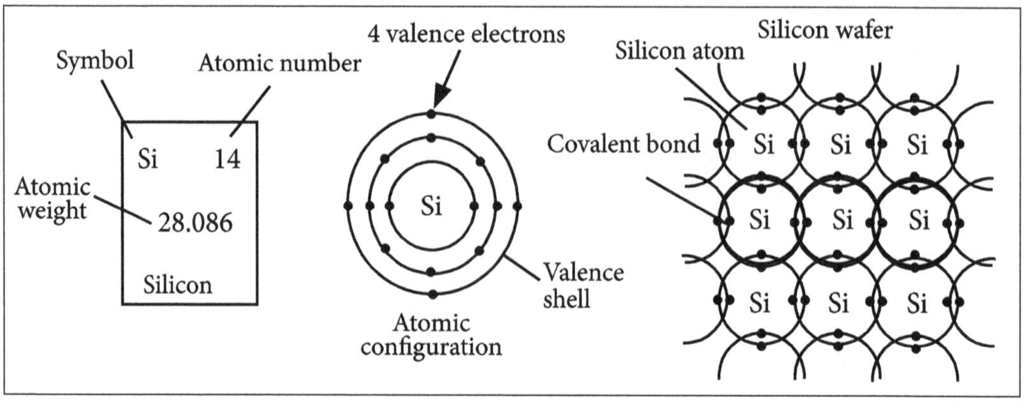

N-type Silicon

Pentavalent impurities such as phosphorus, arsenic, antimony and bismuth have five valence electrons. When phosphorus impurity is added to Si, every phosphorus atoms's four valence electrons are locked up in the covalent bond with the valence electrons of four neighbouring Si atoms. However, the 5th valence electron of the phosphorus atom does not find the binding electron and hence it remains free to float. When an voltage is applied across the silicon-phosphorus mixture, free electrons migrate towards the positive voltage end.

When the phosphorus is added to Si to yield the above effect, we say that Si is doped with phosphorus. The resulting mixture is referred as N-type silicon. The pentavalent impurities are referred to as donor impurities.

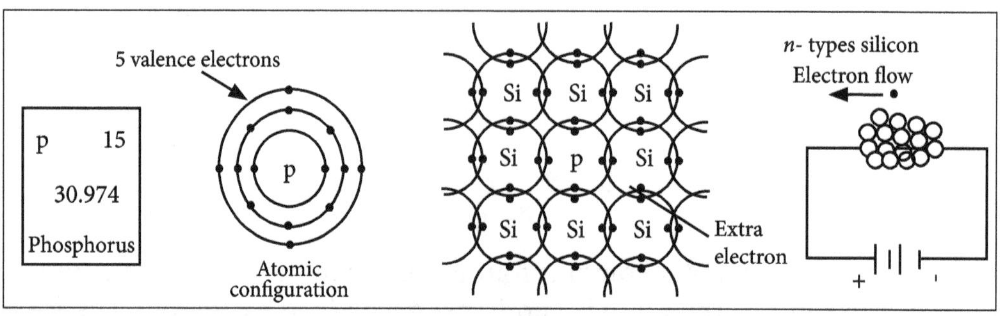

N-type silicon.

P-type Silicon

Trivalent impurities such as boron, aluminum, indium and gallium have three valence electrons. When boron is added to Si, every boron atom's three valence electrons are locked up in the covalent bond with the valence electrons of three neighbouring Si atoms. However, the vacant spot hole is created within the covalent bond between one boron atom and a neighbouring Si atom. The holes are considered as positive charge carriers.

When a voltage is applied across the silicon-boron mixture, the hole move towards the negative voltage end while the neighboring electron fills in its place. When boron is added to Si to yield the above effect, we say that Si is doped with boron. The resulting mixture is termed as P-type silicon. The trivalent impurities are referred to as acceptor impurities.

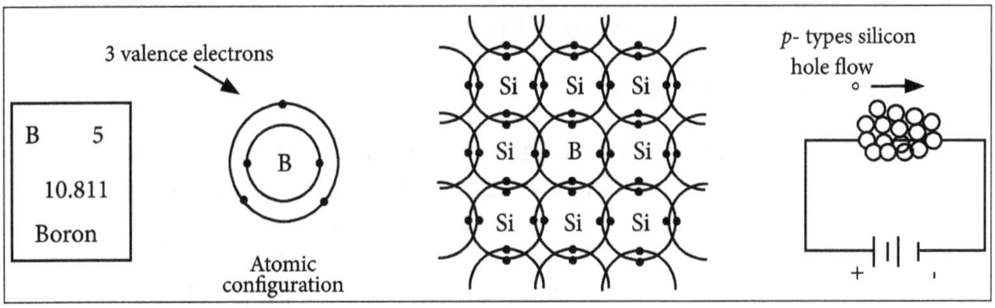

P-type silicon.

The hole of the boron atom points towards the negative terminal. The electron of the neighbouring silicon atom points towards the positive terminal. The electron from neighbouring silicon atom falls into the boron atom thus, filling the hole in the boron atom and creating a new hole in the silicon atom. It appears as though a hole moves towards the negative terminal.

PN Junction Diode Act as Rectifier

A PN Junction diode is a two terminal device that is polarity sensitivity. When the diode is forward biased, the diode conducts and allows current to flow through it without any resistance. (i.e.) the diode is ON.

When the diode is reverse biased, the diode does not conduct and no current flows through it (i.e.) the diode is OFF or providing the blocking function. Thus, an ideal diode acts as a switch; either opened or closed depending upon the polarity of the voltage applied across it. An ideal diode has zero resistance under forward bias and infinite resistance under reverse bias.

Avalanche Breakdown in Transistor

When electrons and holes in semiconductors are accelerated by applied voltage, these accelerated holes and electron collide with the bound electrons and produces free electrons. This causes even more collision due to increase in current value. This is called as avalanche breakdown.

Transition Capacitance of the Diode

A reverse biased condition capacitance exists at the PN junction, where the two regions act as plates and the depletion region acts as dielectric. This is called as transition capacitance of the diode.

Reverse Saturation Current

The current due to minority carriers in reverse bias is said to be reverse saturation current. This current depends on the value of the reverse bias voltage.

p-n Junction Diode as Junction Transistor

A Transistor as Two Diodes

Transistors are like an extension of another semiconductor component such as diodes. In a way transistors are just two diodes with their cathodes or anodes tied together.

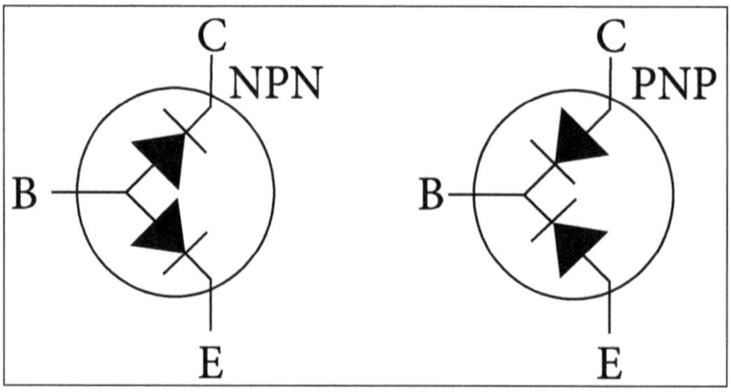

The diode connecting the base to the emitter is the important one. It matches the direction of the arrow on the schematic symbol and shows us which way current is intended to flow through the transistor.

5.4 Insulators: Electrical and Electronic Applications

Insulators

They do not practically allow the heat or electricity to flow through them. Example, most of the organic and inorganic solids, except graphite.

Electrical insulator

Electrical insulators are used in electrical system to prevent unwanted flow of current to the earth from its supporting points. The insulator plays a vital role in electrical systems. Electrical insulator is a very high resistive path through which practically no current can flow. In transmission and distribution system, the overhead conductors are normally supported by supporting towers or poles. Both the towers and poles are properly grounded. So there must be a insulator between tower or pole body and current carrying conductors to prevent the flow of current from conductor to the earth through grounded supporting towers or poles.

Electrical and Electronic Applications

An electrical insulator, also known as a non-conductor, is used to prevent the flow of electric currents. Materials that work as non-conductors lack movable electric charges. Highly resistant electrical insulators includes mica, glass, quartz and solid ionic substances. These materials work by tightly binding electrons so there is no electron flow between them and therefore, no conduction. Electrical insulators may be used in high voltage systems, circuit boards and as coating on electric wires and cables.

Generally, fine ceramics also known as advanced ceramics are insulating materials that do not conduct electricity. A few examples of products that utilize the insulation property of fine ceramics include packages for surface mounted electronic components, such as quartz crystal oscillators and Surface Acoustic Wave filters. These products are widely used in mobile phones, automotive navigation systems and portable music players. Ceramic packages provides advanced hermetic sealing and electrical insulation between electric circuit lines to maintain high reliability of these electronic components.

5.5 Magnetic Materials: Ferro and Ferrimagnetism

Magnetization is defined as net dipole moment per unit volume.

Magnetization vector,

$$M = \frac{Q_m l}{Al}.$$

Difference Between Hard and Soft Magnetic Materials

S. no.	Hard magnetic materials	Soft magnetic materials
1.	Cannot be magnetized easily.	Can be easily magnetized.
2.	It can be produced by heating and sudden cooling.	It can be produced by heating and slow cooling.
3.	Domain walls does not move easily and require large value of H for magnetization.	Domain walls moves easily and requires small value of H for magnetization.
4.	Hysteresis loop area is large.	Hysteresis loop area is small.
5.	Retentivity and coercivity are large.	Coercivity and retentivity are small.
6.	High eddy current loss.	Low eddy current loss.
7.	Impurities and defects will be more.	No impurities and defects.
8.	Examples: Alnico, chromium steel, tungsten steel, carbon steel.	Examples: Iron silicon alloy, ferrous nickel alloy, ferrites garnets.
9.	Susceptibility and permeability values are small.	Susceptibility and permeability values are high.
10.	Uses: Permanent magnets, DC magnets.	Uses: Electromagnets, computer data storage. Transformer core.

Domain Theory of Ferromagnetism

This theory was proposed by Weiss in 1907. It explains the hysteresis and the properties of ferromagnetic materials.

Magnetic Domains

A ferromagnetic material is divided into a large number of small region called domains, each direction is spontaneously magnetized. The direction of magnetization varies from domain to domain and the net magnetization is zero, in the absence of external magnetic field.

The boundary line which separates two domains is called domain wall or block wall. When the magnetic field is applied to the ferromagnetic material, the magnetization is produced by two ways as:

- By the motion of domain walls.
- By the rotation of domains.

Process of Domain Magnetization

1. By the Motion of Domain Walls

When a small amount of magnetic field is applied, the domains having dipoles parallel to the applied magnetic field increases in area by the motion of domain walls.

2. By the Rotation of Domains

If the applied magnetic field is further increased, the domains are rotated parallel to the field direction by the rotation of domains.

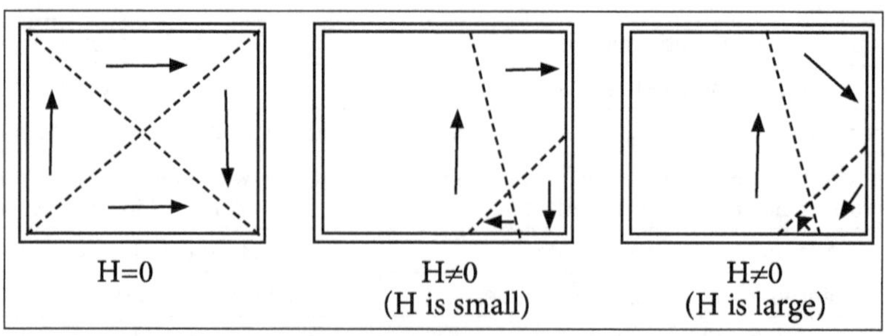

Domain theory of ferromagnetism.

Origin of Domain Theory of Ferromagnetism or Energies Involved in the Domain Growth

The origin of domains can be understood from the thermodynamic principle (i.e.) in equilibrium, the total energy of the system is minimum.

The total internal energy of the domain structure in a ferromagnetic material is made up from the following contributions:

- Exchange energy or magnetic field energy.
- Crystalline energy or anisotropy energy.
- Domain wall energy or bloch wall energy.
- Magnetostriction energy.

1. Exchange Energy or Magnetic Field Energy

The interaction energy which makes the adjacent dipoles align themselves is called as exchange energy or magnetic field energy. It arises from the interaction of electron spins. It depends on the inter atomic distance. This exchange energy also called magnetic field energy is the energy required in assembling the atomic magnets into a single domain and this work done is stored as potential energy.

The size of the domains for a particular domain structure may be obtained from the principle of minimum energy. The volume of the domain may vary between, 10^{-2} to 10^{-6} cm^3.

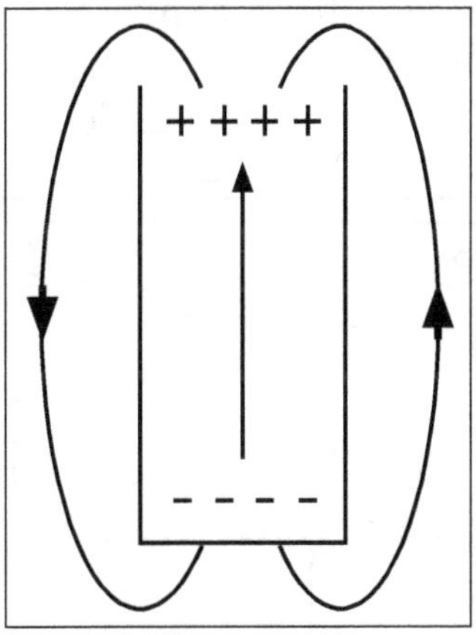

Single domain.

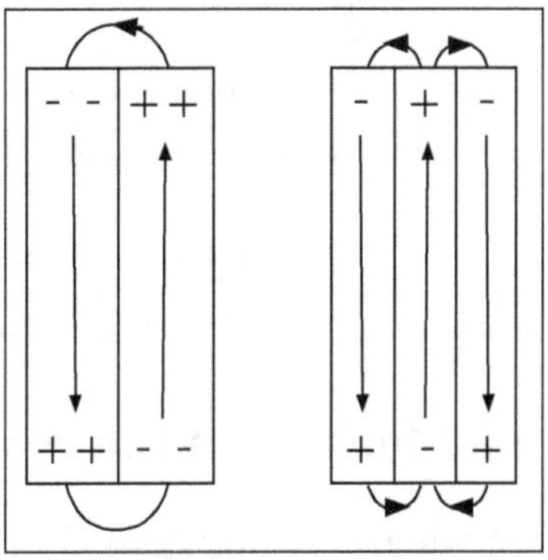

Multi domain.
Exchange energy in ferromagnetism.

2. Crystalline Energy or Anisotropy Energy

The excess energy required to magnetize a specimen in particular direction over that required to magnetize it along the easy direction is called the crystalline anisotropy energy.

In ferromagnetic materials there are two types of directions of magnetization namely,

- Easy direction.
- Hard direction.

In easy direction of magnetization, weak fields can be applied and in hard direction of magnetization, strong fields should be applied.

Crystalline anisotropy energy is the energy of magnetization which is the function of crystal orientation. As shown in figure, magnetization curves for iron with the applied field along different crystallographic directions have been drawn.

For example, in BCC iron the easy direction is [100], the medium direction is [110] and the hard direction [111]. The energy difference between hard and easy direction to magnetize the material is about magnetic anisotropy energy. This energy is very important in determining the characteristic domain boundaries.

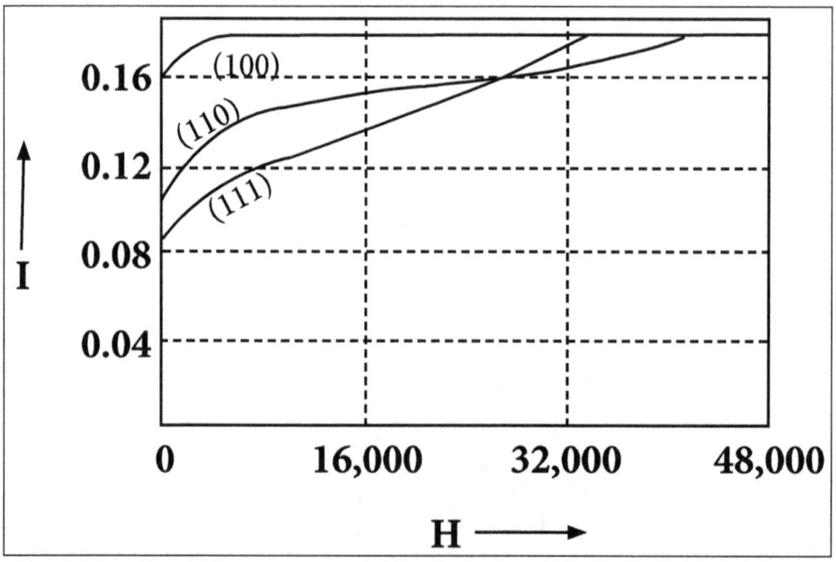

Anisotropy energy in ferromagnetism.

3. Domain Wall Energy or Bloch Wall Energy

A thin boundary or region that separates adjacent domains magnetized in different directions is called domain wall or Bloch wall.

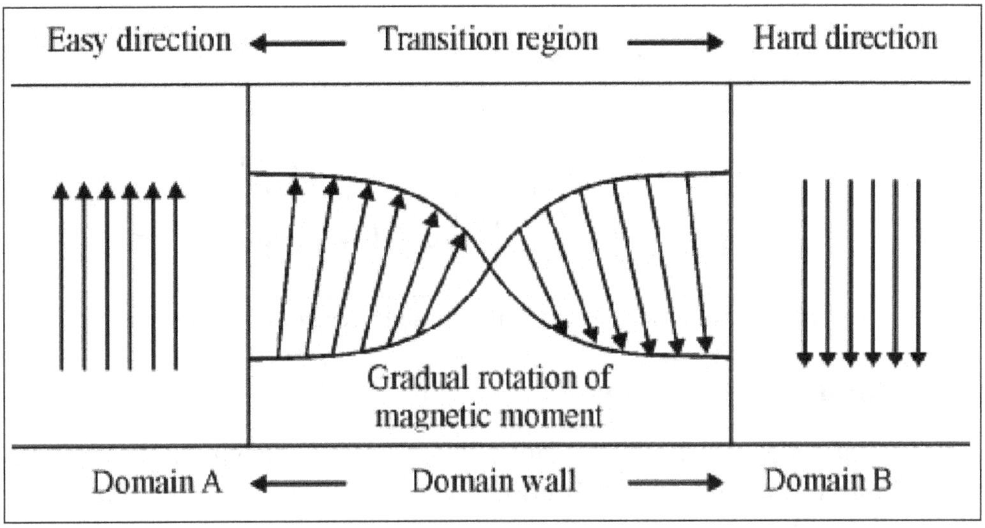

The change of electron spin in the transition region of Bloch wall.

The size of the Bloch wall is about 200 to 300 lattice constant thickness. In going from one domain to another domain, the electron spin changes gradually as shown in figure. The energy of domain wall is due to both exchange energy and anisotropic energy.

Based on the spin alignments, two types of Bloch walls may arise namely,

I. Thick Wall

When the spins at the boundary are misaligned and if the direction of the spin changes gradually as shown in figure, it leads to a thick Bloch wall. Here the misalignments of spins are associated with exchange energy.

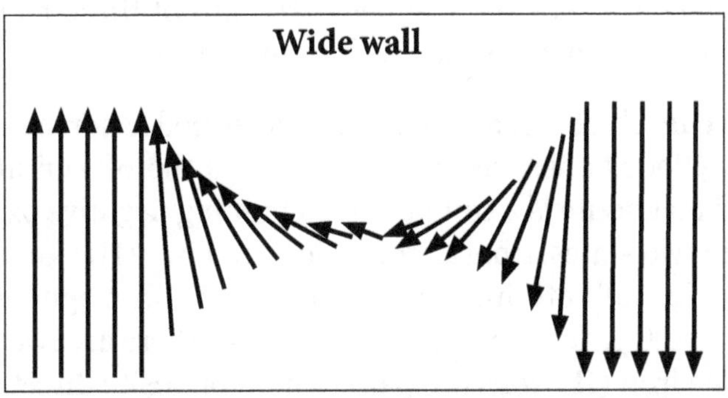

The change of electron spin in the transition region of thick wall.

II. Thin Wall

When the spins at the boundaries changes abruptly, then the anisotropic energy becomes very less. Since the anisotropic energy is directly proportional to the thickness of the wall, this leads to a thin Bloch wall.

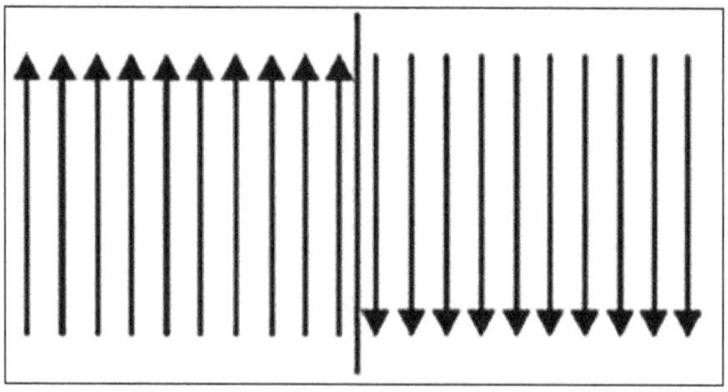

The change of electron spin in the transition region of thin wall.

4. Magnetostriction Energy

When a material is magnetized, it is found that there is a change in dimension. This phenomenon is known as magnetostriction. This deformation is different along different crystal directions.

Hence, if the domains are magnetized in different directions, they will either shrink or expand. This means that work should be done against the elastic restoring forces. The work done by the magnetic field against these elastic restoring forces is known as magneto-elastic energy or magnetostrictive energy.

Ferrimagnetism

Ferrimagnetism is a type of permanent magnetism that occurs in solids in which the magnetic fields associated with individual atoms spontaneously align themselves, some parallel or in the same direction and others generally antiparallel or paired off in opposite directions. The magnetic behaviour of single crystals of ferrimagnetic materials may be attributed to the parallel alignment, the diluting effect of those atoms in the antiparallel arrangement keeps the magnetic strength of these materials usually less than that of purely ferromagnetic solids such as metallic iron.

Ferrimagnetism occurs chiefly in magnetic oxides termed as ferrites. The natural magnetism exhibited by lodestones, recorded as early as the 6th century BC, is that of a ferrite, the mineral magnetite is a compound containing negative oxygen ions O^{2-} and positive iron ions in two states, iron(II) ions, Fe^{2+} and iron(III) ions, Fe^{3+}. The oxygen ions are not magnetic, but both iron ions are magnetic. In magnetite crystals, chemically formulated as Fe_3O_4, for every four oxygen ions, there are two iron (III) ions and one iron(II) ion. The iron (III) ions are paired off in opposite directions, producing no external magnetic field, but the iron (II) ions are all aligned in the same direction, accounting for the external magnetism.

The spontaneous alignment that produces ferrimagnetism is entirely disrupted above a temperature called Curie point, characteristic of each ferrimagnetic material. When the temperature of the material is brought below the Curie point, ferrimagnetism revives.

5.5.1 Hall Effect and its Applications

Hall Effect

When conductor carrying a current is placed in a transverse magnetic field, an electric field is generated inside the conductor in a direction normal to both the current and the magnetic field. This phenomenon is termed as Hall effect and the generated voltage is called Hall voltage.

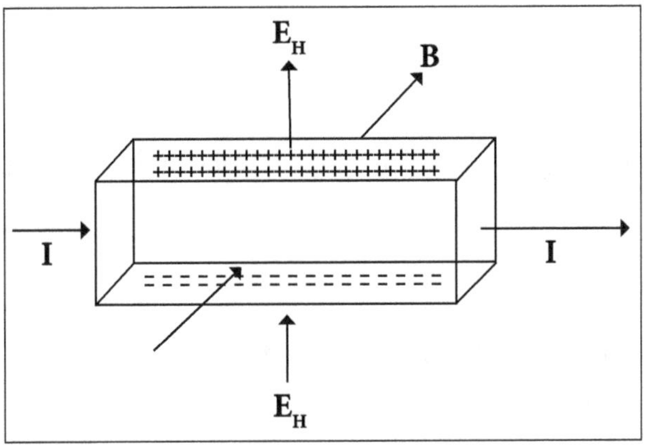

Hall effect.

Hall Effect in n-type Semiconductor

Let us consider an n-type material into which the current is allowed to pass along the x-direction from left to right and the magnetic field is applied in z-directions, as a result Hall voltage is produced in y-direction.

Since the direction of current is from left to right, the electrons moves from right to left in x-direction as shown in the figure.

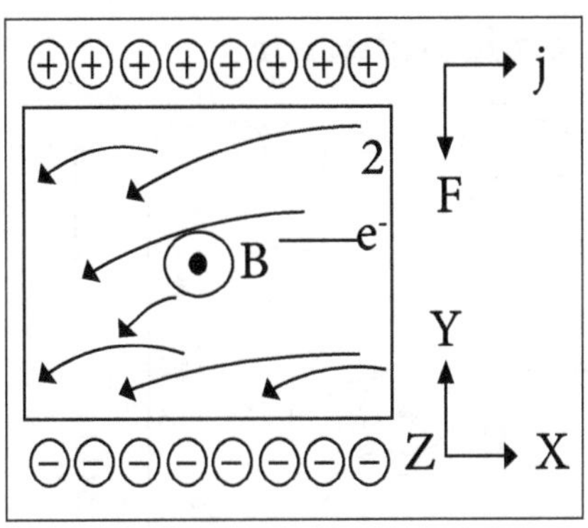

Hall effect in n-type semiconductor.

Now due to the magnetic field applied, the electrons move towards downward direction with the velocity v and cause the negative charge to accumulate at face (1) of the material as shown in the figure.

Therefore, a potential difference is established between face (2) and face (1) of the specimen which gives rise to field E_H in the negative y-direction.

Here, the force due to potential difference $= -eE_H$ ...(1)

Force due to magnetic field $= -Bev$ ...(2)

At equilibrium equation (1) = equation (2).

$$-eE_H = -Bev$$

$$E_H = Bv \quad ...(3)$$

We know that, the current density J_x in the x-direction is, $J_x = -n_e ev$

$$V = \frac{J_X}{N_e e} \quad ...(4)$$

Substituting equation (4) in equation (3) we get,

$$E_H = \frac{BJ_X}{n_e e} \quad ...(5)$$

$$E_H = R_H J_x B \quad ...(6)$$

Where R_H is known as the Hall co-efficient, which is given by,

$$R_H = -(1/n_e e) \quad ...(7)$$

The negative sign indicates that the field is developed in the negative y-direction.

Hall Effect in p-type Semiconductor

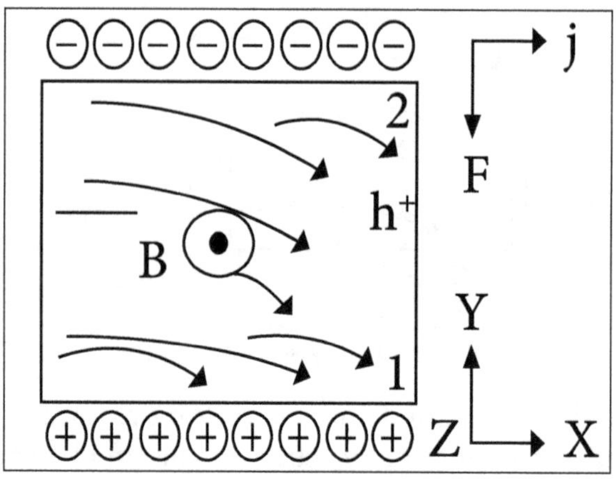

Hall effect in p-type semiconductor.

Let us consider a p-type material for which the current is passed along the x-direction from left to right and magnetic field is applied along z-direction as shown in the figure. Since the direction of current is from left to right, the holes will also move in the same direction.

Now due to the applied magnetic field, the holes move towards the downward direction with velocity v and accumulate at the face (1) as shown in the figure.

A potential difference is established between face (1) and (2) in the positive y-direction.

Force due to the potential difference $= eE_H$ ...(8)

At equilibrium, equation (7) = equation (8).

$$eE_H = Bev$$

$$E_H = Bv$$

We know that, the current density,

$$J_x = n_h ev \quad ...(9)$$

$$V = \frac{J_X}{n_h} e \quad ...(10)$$

Where, n_h is the hole density.

Substituting equation (10) in (9) we get,

$$E_H = \frac{BJ_X}{n_h e}$$

$$E_H = R_H J_x B \quad ...(11)$$

Where,

$$R_H = \frac{1}{n_h e}$$

Equation (11) represents the hall co-efficient and the positive sign indicates that the hall field is developed in the positive y-direction.

Experimental Determination of Hall Effect

A semiconductor slab of thickness 't' and breadth 'b' is taken and current is passed using a battery as shown in the figure.

The slab is placed between the pole of an electromagnet so that the current direction coincides with x-axis and the magnetic field coincides with z-axis. The hall voltage (V_H) is measured by placing two probes at the center of the top and bottom faces of the slab.

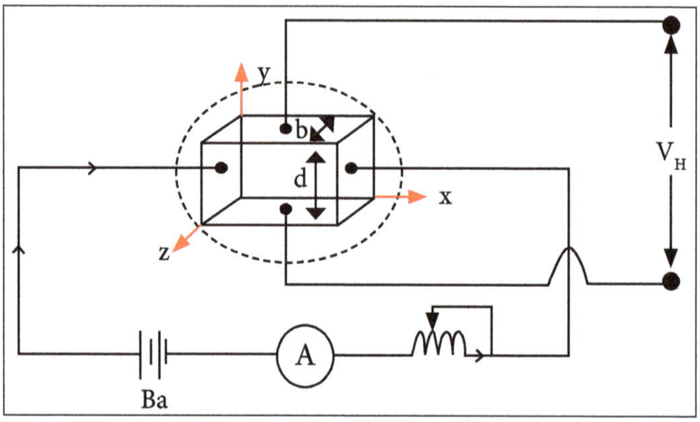

Experimental setup for Hall effect.

If B is the magnetic field applied and the VH is the Hall voltage produced, then the Hall co-efficient can be calculated from the formula,

$$R_H = \frac{V_H b}{I_X B} \qquad ...(12)$$

Mobility of charge carriers

In general, the hall co-efficient can be written as,

$$R_H = \frac{-1}{ne} \qquad ...(13)$$

The above expression is valid only for conductors where the velocity is taken as the drift velocity. But for semiconductors, velocity is taken as average velocity. Hence, RH for an n type semiconductor is modified as,

$$R_H = \frac{-3\pi}{8}\left[\frac{1}{n_e e}\right]$$

$$R_H = \frac{-1.18}{n_e e} \qquad ...(14)$$

We know that, the conductivity for n type is $\sigma = n e \mu_e$

$$\mu_e = \frac{\sigma_e}{n_e e} \qquad ...(15)$$

Equation (3) can be rewritten as,

$$\frac{1}{n_e e} = \frac{-R_H}{1.18} \quad ...(16)$$

Substituting equation (5) in (4) we get,

$$\mu_e \frac{-\sigma_e R_H}{1.18} \quad ...(17)$$

The mobility of electron in an n-type semiconductor can be written as,

$$\mu_e \frac{\sigma_e V_H b}{1.18\, I \times B} \quad \left[\because R_H = \frac{V_H b}{I \times B}\right]$$

Similarly for p-type semiconductor, the mobility of hole can be written as,

$$\mu_e = \frac{\sigma_h V_H b}{1.18\,(I \times B)} \quad ...(18)$$

Thus, by finding hall voltage, hall co-efficient can be calculated and the mobility of charge carriers can also be determined.

Applications of Hall Effect

- The sign of charge carriers can be determined.
- The carrier concentration can be determined. $\left[n = \frac{1.18}{qR_H}\right]$.
- The mobility of charge carriers is measured directly. $\left[\mu = \frac{-\sigma R_H}{1.18}\right]$.
- Electrical conductivity can be determined $\sigma = nq\mu$.
- It can be used to determine whether the given material is metal, insulator or semiconductor and the type of the semiconductor.
- It can be used to determine the power flow in an electromagnetic wave.

6

Non-Conventional Energy Sources and Storage Devices

6.1 Application and Conversion of Solar Energy

Solar Energy

Energy that we get directly from the sun is called as solar energy.

The nuclear fusion reactions occurring inside sun release enormous amount of energy in the form of heat and light. Several techniques are available for collecting, converting and using solar energy.

Methods of Harvesting Solar Energy

Some important solar energy harvesting devices are given below:

1. Solar Battery

When a large number of solar cells are connected in series it form a solar battery. The solar battery produces electricity capable of running water pump to run street-light etc. They are used in remote areas where the conventional electricity supply is a problem.

2. Solar Heat Collectors

Solar pump run by solar cells.

Solar heat collectors consists of natural materials like stones, bricks (or) materials like glass, which can absorb heat during the day time and release it slowly at night.

Uses

It is generally used in cold places, where houses are kept in hot condition using the solar heat collectors.

3. Solar Water Heater

It consists of an insulated box inside of which is painted with black paint. It is also provided with a glass lid to receive and store solar heat. Inside the box, it has a black painted copper coil, through which cold water is allowed to flow in, which gets heated up and flows out into a storage tank. From the storage tank water is then supplied through pipes.

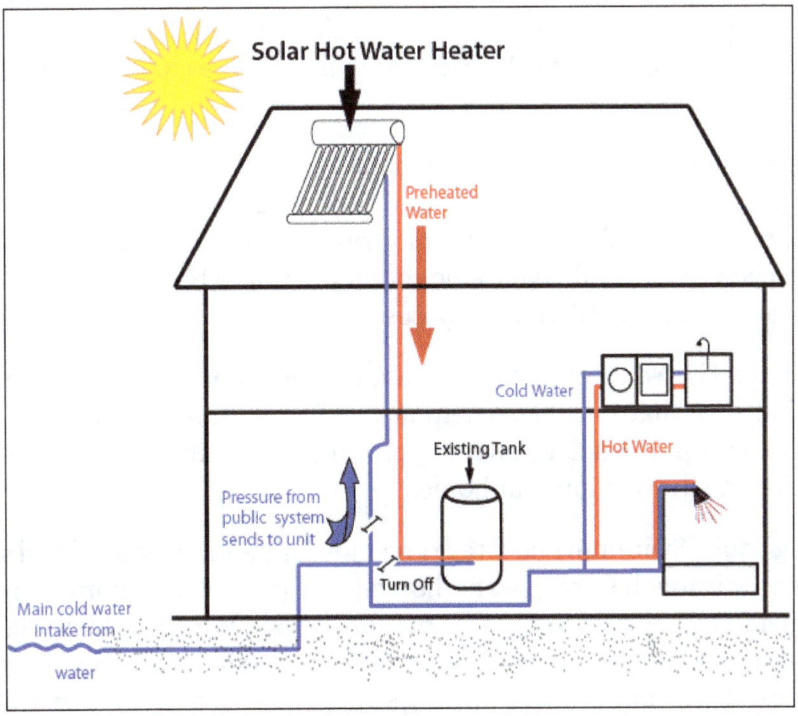

Solar water heater.

Significance of Solar Energy

- Solar cells are free from air and sound pollution.
- Solar water heaters, cookers, require neither fuel not attention while cooking food.
- Solar cells can be used in remote and isolated areas, forests, hilly regions.

Applications of Solar Energy

1. Power plants: In conventional power plants non-renewable energy sources are used to boil water and form stream so that turbines can rotate and water to produce electricity. But with application of the solar energy heat of sun can boil that water to create steam and rotate turbines. To convert sunlight into electricity solar panels, photoelectric technologies and the thermoelectric technologies etc. are used.

2. Homes: Use of solar energy is increasing in homes as well. The residential appliances can easily use electricity generated through solar power. Besides this solar energy is running solar heater to supply hot water in homes. Through photovoltaic cell installed on the roof of the house energy is captured and stored on batteries to use throughout the day at homes for different purposes. In this ways expenditure on the energy is cutting down by home users.

3. Commercial use: On roofs of different buildings we can find glass PV modules or any other kind of solar panel. These panels are used there to supply electricity to different offices or other parts of the building in a reliable manner. These panels collect solar energy from sun, convert it into electricity and allow offices to use their own electrical power for different purposes.

4. Ventilation system: At many places solar energy is used for the ventilation purposes. It helps in running bath fans, floor fans and ceiling fans in buildings. The fans run almost every time in a building to control moisture and smell and in homes to take heat out of the kitchen. It can add heavy amount on the utility bills, to cut down these bills solar energy is used for ventilation purposes.

5. Power pump: The Solar power is not just help in improving ventilation system at our homes but with that it can also help in circulating water in any building. We can connect power pump with solar power supply unit but we must run it on DC current so that water circulate throughout our home.

6. Swimming pools: Swimming pools are great joy for kids and adults in all seasons. But during winters it is tough to keep water hot in these pools with minimum power usage. The solar energy can help many in this matter as well. We can add a solar blanket in the pool that will keep the water hot with energy generated from the sunlight. Besides this we can install a solar hot water heating system with solar hot water heating panels.

7. Solar lighting: It also known as day lighting and work with help of solar power. It store natural energy of the sun in day time and then converts this energy into electricity to light up in night time. Use of this system is reducing load form local power plants.

8. Solar cars: It is an electrical vehicle which is recharged form the solar energy or sunlight. Solar panels are used on this car that absorb light and then convert it into electrical energy. This electrical energy is stored in batteries used with the car, so that in night time also we can drive these vehicles.

9. Remote applications: Remote buildings are taking benefit of solar energy at vast scale. Remote schools, community halls and clinics can take solar panel and batteries with them anywhere to produce and use electric power.

Conversion of Solar Energy

1. Solar Thermal Conversion

Major Functions:

- Solar Radiation collection and concentration.
- Conversion to heat.
- Storage of energy.
- Generation of electricity.

Solar energy is collected as high-temperature heat, normally by means of mirrors or lenses that track the motion of the sun and direct a concentrated solar flux onto a receiver. Temperatures up to 1000K can be generated by this means, high enough to produce the high-pressure steam used in the modern steam turbines to generate electricity.

Collection

The temperature to which a surface is heated by certain flux of incident solar energy is determined by balance of incident radiation and loss by conduction, convection and radiation. Use of selective surfaces that absorb visible sunlight but do not lose energy by infrared radiation will achieve high temperatures.

Temperature obtained can be increased by boosting the flux of incident sunlight by the use of concentrating mirrors or lenses. A fairly low concentration ratio, obtainable with simple optics, can be combined with a selective surface to efficiently produce temperatures high enough for electrical power generation.

Three basic collection geometries of the sunlight for solar thermal conversion:

- Non-concentrating.
- Concentrating to a line.
- Concentrating to a point.

Non-Concentrating Collectors

Industrial process heat at moderate temperatures (About 18 percent of the fuel consumption in the United States is for generation of industrial process heat at moderate temperatures (process steam) and another 11.5 percent is used for high-temperature process heat). Low temperature heat for domestic water heating.

Concentration to a Line

The concentration ratios up to 20 with selective surfaces to achieve high enough temperatures for the electrical generation.

Point Focusing

Concentration ratios up to 1000, without the need for use of selective surfaces. (Thermochemical conversion - splitting of water to produce hydrogen).

Solar Power Tower

A solar power tower or central receiver generates electricity from sunlight by focusing concentrated solar energy on a tower-mounted heat exchanger (receiver). This system uses hundreds to thousands of flat, sun-tracking mirrors called heliostats to reflect and concentrate sun's energy onto a central receiver tower. The energy can be concentrated as much as 1,500 times that of the energy coming in from the sun.

Energy losses from thermal-energy transport are minimized because solar energy is being directly transferred by the reflection from the heliostats to a single receiver, rather than being moved through a transfer medium to one central location, as with parabolic troughs.

Power towers must be large to be economical. This is promising technology for large-scale grid-connected power plants.

The application of solar energy by thermal conversion is used for:

- Sanitary water heating in homes, apartments, hotels, hostels, student dormitories, homes for elderly, kindergartens, restaurants and sports facilities.
- Heating of sanitary water for villages that have a distribution of hot water from the urban heating plants in the period when heating plants are not working.
- Heating pools in homes and sports centers.
- Heating water or other fluids in industrial processes.
- Heating greenhouses in agricultural production.
- Drying agricultural products.
- Distillation of water for industrial purposes.
- Heating rooms as an additional tool in the periods when there are enough sunny days.
- Production of electricity based on the thermal conversion of solar radiation (steam turbines).

- In the processes for room cooling.

2. Photo Conversion

The amount of energy striking earth from sunlight in one hour is equal to all the energy consumed globally from fossil fuels in a year. Thus the solar flux provides an enormous source of renewable energy, however, its economic use requires a stable method to store and distribute energy. The only scalable and efficient method to store solar energy is by its conversion into chemical bonds. Thus, the use of sunlight to split water or artificial photosynthesis is Holy Grail of renewable energy research.

We are pursuing a water-splitting photo conversion system that will produce hydrogen fuel directly from the water using only inexpensive materials. Conceptually our cell is built around a semiconductor membrane that converts the sunlight into electron/hole pairs. The photo generated electrons and holes would be directed to opposite sides of membrane where catalysts in contact with a water solution would produce hydrogen and oxygen. Fundamentally the materials must be able to generate the 2.46eV of free energy needed to split water into H_2 and 1/2 O_2.

The cell will be constructed of two semiconductors as light absorbers embedded in an electrical and ion conduction membrane. Membrane will separate the cell into two compartments where either the water oxidation or reduction reactions will occur. Absorbers will be made from cheap polycrystalline semiconductors and will separate the electrons and holes across the liquid/semiconductor junction. In order to achieve efficient charge separation the semiconductors will be formed into a micro/nano-rod assembly with the catalysts for water reduction and oxidation attached to the sides of the semiconductors. Oxidation and reduction will occur on opposite sides of the membrane, thus requiring each semiconductor to be robust to a single reaction and use of two different catalysts. The system will make use of two solar photons to provide the energy to drive the reactions. The development of a solar photo conversion device will require catalyst develop studies of kinetics and mechanism of hydrogen and oxygen generation, knowledge of how to attach active catalysts to semiconductor surfaces, semiconductor development as well as membrane construction.

6.2 Photovoltaic Cell: Design, Working and its Importance

A device which converts the solar energy (energy obtained from sun) directly into electrical energy is called 'as Solar cell'. This is also called as 'Photovoltaic cell'.

Principle

The basic principle involved in the solar cells is based on the photovoltaic (PV) effect. When sun rays fall on the two layers of the semiconductor devices, potential difference between the two layers is produced. This potential difference causes flow of electrons and thus produces electricity.

Example: Silicon solar cell

Design: Modules, Panels and Arrays of Photovoltaic Cells

A number of photovoltaic cells connected to each other and mounted in a support structure or frame is called photovoltaic module.

One or more modules assembled as pre-wired, field install-able unit produce photovoltaic panels. The complete power generating unit consisting of any number of photovoltaic modules and panels is called photovoltaic arrays.

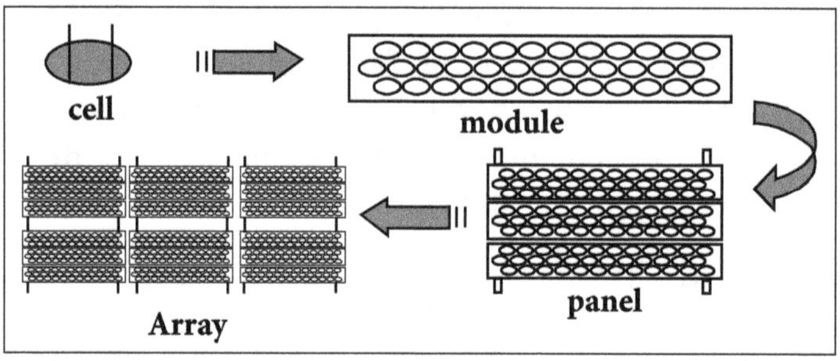

Construction

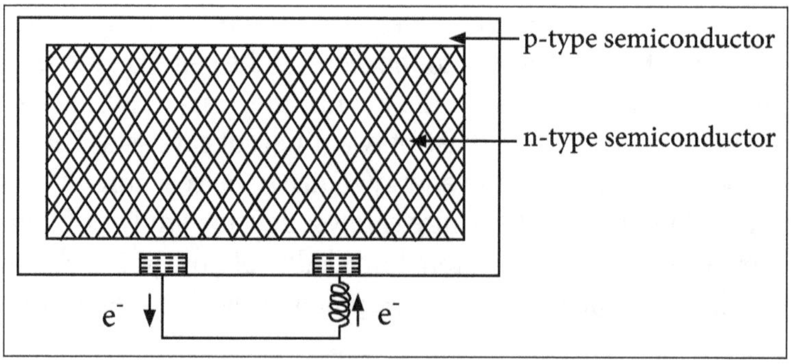

Solar cell consists of a p-type (such as Si doped with boron) and a n-type (such as Si doped with phosphorous).

They are in close contact with each other.

Working

When the solar rays fall on top layer of the p-type semiconductor, the electrons from the valence band get promoted to the conduction band and cross the p-n junction into the n-type semiconductor. Thereby potential difference between two layers gets created, which causes flow of electrons (i.e. electric current).

The potential difference and so current increases as more solar rays falls on the surface of top layer. Thus, when this p- and n- layers are connected to an external circuit, electrons flow from n-layer to p-layer and hence current is generated.

Importance

- PV panels provide clean – green energy. During electricity generation with PV panels there is no harmful greenhouse gas emissions, thus solar PV is environmentally friendly.
- The solar energy is energy supplied by nature – it is thus free and abundant.
- The solar energy can be made available almost anywhere there is sunlight.
- The solar energy is particularly appropriate for smart energy networks with distributed power generation – DPG is indeed the next generation power network structure.
- Solar Panels cost is currently on a fast reducing track and is expected to continue reducing for the next years – consequently solar PV panels has indeed a highly promising future both for economic viability and environmental sustainability.
- Photovoltaic panels, through the photoelectric phenomenon, produce electricity in a direct electricity generation way.
- Operating and maintenance costs for the PV panels are considered to be low, almost negligible, compared to costs of other renewable energy systems.
- PV panels have no mechanically moving parts, except in cases of –sun-tracking mechanical bases, consequently they have far less breakages or require less maintenance than other renewable energy systems (e.g. wind turbines).
- PV panels are totally silent, producing no noise at all; consequently, they are a perfect solution for urban areas and for residential applications.

6.3 Non-Conventional Energy Sources: Hydropower Include Setup a Hydropower Plant

Non-Renewable energy sources are limited in their availability. Acquisition and usage of Non-Renewable energy sources causes disruption and disturbs the balance of environment.

Non-Renewable Energy Sources

- Coal

- Nuclear Energy
- Oil
- Natural Gas

Hydropower Plant

Power system mainly contains three parts namely generation, transmission and distribution. Generation means how to generate the electricity from the available source and there are various methods to generate electricity but here we only focused on generation of electricity by the means of hydro or water (hydro power plant).

In hydro power plant we use gravitational force of fluid water to run the turbine which is coupled with electric generator to produce electricity. It plays an important role to protect our fossil fuel which is limited, because the generated electricity in hydro power station by the use of water which is the renewable source of energy and it is available in a large amount without any cost.

The advantage of hydro power is the water which the main stuff to produce electricity in hydro power plant is free, it not contain any type of pollution and after generated electricity the price of electricity is average not too much high.

Essential Elements

1. Dam

The dam is the most important component of hydroelectric power plant. Dam is built on a large river that has abundant quantity of water throughout the year. It should be built at a location where the height of the river is sufficient to get the maximum possible potential energy from water.

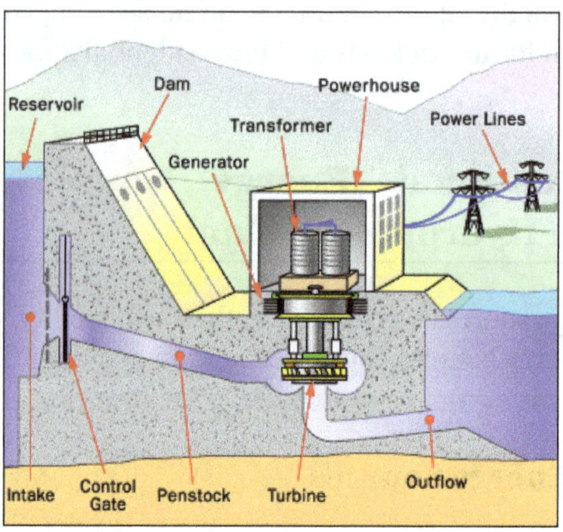

Hydro Power plant- Essential elements.

2. Water Reservoir

The water reservoir is the place behind the dam where water is stored. Water in the reservoir is located higher than the rest of the dam structure. The height of water in the reservoir decides how much potential energy the water possesses. The higher the height of water, the more its potential energy. The high position of water in the reservoir also enables it to move downwards effortlessly.

The height of water in reservoir is higher than the natural height of water flowing in the river, so it is considered to have an altered equilibrium. This also helps to increase the overall potential energy of water, which helps ultimately produce more electricity in power generation unit.

3. Intake or Control Gates

It is built on the inside of the dam. The water from reservoir is released and controlled through these gates. These are called inlet gates because water enters the power generation unit through these gates. When the control gates are opened the water flows due to gravity through the penstock and towards the turbines. The water flowing through the gates possesses potential as well as kinetic energy.

4. The Penstock

It is the long pipe or the shaft which carries the water flowing from the reservoir towards the power generation unit, comprised of the turbines and generator. The water in the penstock possesses the kinetic energy due to its motion and potential energy due to its height.

The total amount of power generated in the hydroelectric power plant depends on the height of the water reservoir and the amount of water flowing through the penstock. The amount of water flowing through the penstock is controlled by the control gates.

5. Water Turbines

Water flowing from the penstock is allowed to enter the power generation unit, which houses the turbine and generator. When water falls on the blades of the turbine the kinetic and potential energy of water is converted into rotational motion of the blades of the turbine. The rotating blades causes the shaft of the turbine to also rotate. Turbine shaft is enclosed inside the generator. In most hydroelectric power plants there is more than one power generation unit.

There is a large difference in the height between the level of the turbine and the level of water in the reservoir. This difference in height is also known as the head of water, which decides the total amount of power that can be generated in hydroelectric power plant.

There are various types of water turbines such as Kaplan turbine, Francis turbine, Pelton wheels etc. The type of turbine used in the hydroelectric power plant depends on the height of the reservoir, quantity of water and the total power generation capacity.

6. Generators

It is in the generator where electricity is produced. The shaft of the water turbine rotates in the generator, which produces alternating current in the coils of generator. It is the rotation of the shaft inside the generator that produces magnetic field which is converted into electricity by the electromagnetic field induction.

Hence the rotation of the shaft of the turbine is crucial for production of electricity and this is achieved by the kinetic and potential energy of water. Thus in hydroelectricity power plants potential energy of water is converted into electricity.

6.4 Geothermal Energy: Schematic Diagram of a Geothermal Power Plant

The word geothermal comes from the greek words Geo and thermal. It is the heat from within the Earth. It is generated in the earth's core almost 4,000 miles below the Earth's surface. The double layered core is made up of very hot magma surrounding the solid iron center. Very high temperatures are produced continuously inside the earth by slow decay of radioactive materials.

This process is natural in all rocks. Surrounding the outer core is the mantle, which is about 1,800 miles thick and it is made up of magma and rock. The outermost layer of the Earth, the land that forms the continents and ocean floors is called the crust. It is three to five miles thick under the oceans and 15 to 35 miles thick on the continents.

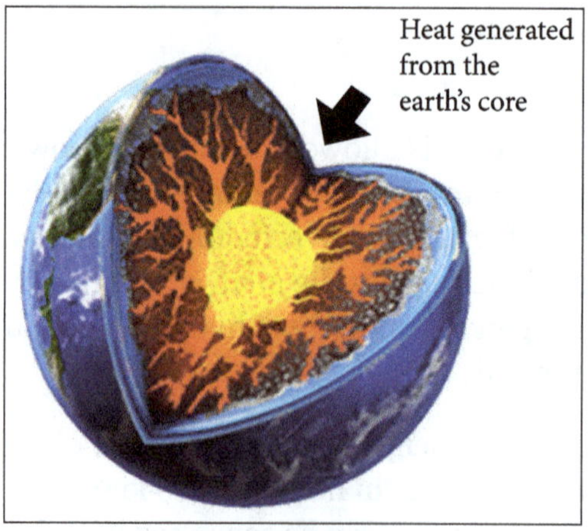

Geothermal energy.

The crust is not a solid, like the shell of an egg but is broken into pieces called as plates.

Magma comes close to the surface of the earth near the edges of these plates. This is the place where volcanoes occur. The lava that erupts from volcanoes is partly magma product. Deep underground the rocks and water absorb the heat from this magma. People around the world use geothermal energy to heat their homes and to produce electricity.

The geothermal energy is a renewable energy source because the water is replenished by rain and the heat is continuously produced deep within the Earth. We won't run out of geothermal energy.

Advantages of Geothermal Energy

Geothermal energy is the energy stored below earth's crust. Due to heat below the earth's crust, water gets heated and sometimes it gets converted into steam. This hot water and steam is used for generating electricity. The advantages of using geothermal energy for producing electricity is:

- Clean source of power: The fuels used in conventional thermal power plants produce lots of toxic gases and particulate matter polluting the surrounding atmosphere. Electricity can be produced from the geothermal reservoir without creating any air pollution.
- No fuel is required: The thermal and nuclear power plants use fossil and nuclear fuels respectively to heat water and generate steam for production of electricity. The geothermal power plants readily uses the available hot water from the geothermal reservoir and hence no fuel is required.
- Unhindered production of power: The energy from the geothermal reservoir is available for 24 hours a day and all the days of the year without any breakage or change during varying seasons, natural disaster and political turmoil. Hence, geothermal energy is the reliable source for producing electricity continuously without any hindrance.
- Less area is required: The land required for geothermal power plant per megawatt of power produced is much lesser than the requirement for the other types of power plants.
- Electrification of the remote places: The geothermal energy reservoir present in the far off and remote areas particularly villages can be utilized for the electrification of these areas. It may be very difficult to supply power to these places by the national grid.
- Additional capacity: The geothermal power plants are very flexible and adding additional capacity for producing more power due to higher demands is quite feasible and affordable.
- Saving money: The fossil fuel used for the thermal power plants has to be imported. By using the geothermal energy lots of local currency is saved.

Disadvantages

Most parts of the world do not have suitable areas where geothermal energy can be exploited.

Geothermal Power Plant

Geothermal electricity is electricity generated from geothermal energy.

All geothermal power plants use steam to turn large turbines, which run electrical generators. In the Geysers Geothermal area, dry steam from below the ground is used directly in the steam turbines. In other areas of the state, super-hot water is "flashed" into steam within the power plant and that steam turns the turbine.

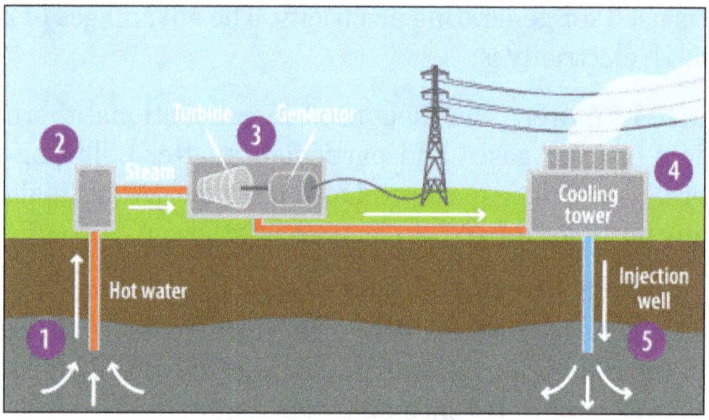

Geothermal Power Plant.

Estimates of the electricity generating potential of geothermal energy vary from 35 to 2000 GW. Current worldwide installed capacity is 10,715 megawatts (MW), with the largest capacity in the United States (3,086 MW), Philippines and Indonesia.

The geothermal power is considered to be sustainable because the heat extraction is small compared with the Earth's heat content. The emission intensity of existing geothermal electric plants is on average 122 kg of CO_2 per megawatt-hour (MW·h) of electricity, about one-eighth of a conventional coal-fired plant.

6.5 Tidal and Wave Power: Design and Working-Movement of Tides and their Effect on Sea Level

Definition of Tidal Power

Tidal power is the power inherent in tides at sea or oceans that is the power of motion of water actuated by tides. The tides are defined as the increase and decrease in water levels due to the motion of water from one place to the other.

This motion of water is actuated by large amounts of energy due to the movement of the Sun, Moon and Earth which are relative to each other and also to their rotational movement. Hence, there is a renewable source of energy in the tidal motion of water at seas and oceans. This source of energy could be used to generate the other types of energy which could be useful in industrial applications.

Wave Power

Wave energy or wave power is essentially power drawn from waves. When the wind blows across sea surface, it transfers the energy to the waves. They are powerful source of energy. Energy output is measured by the wave speed, wave height, and wavelength and water density. The more strong the waves, more capable it is to produce power. The captured energy can then be used for electricity generation, powering plants or pumping of water. It is not easy to harness power from wave generator plants and this is the reason that they are very few wave generator plants around world.

When wind blows across the surface of the water strongly enough it creates waves. This occurs most often and most powerfully on the ocean because of the lack of land to resist the power of wind.

The kinds of waves that are formed depend on from where they are being influenced. Long, steady waves that flow endlessly against beach are likely formed from storms and extreme weather conditions far away. The power of storms and their influence on the surface of the water is so powerful that it can cause waves on the shores of another hemisphere. For example, when Japan was hit with a massive tsunami in 2011, it created powerful waves on the coast of Hawaii and even as far as the beaches of the state of Washington.

Working of Tidal Power Generation Systems

In very simple terms a barrage is built at the entrance of a gulf and the water levels vary on both sides of the small dam. The passages are made inside the dam and water flows through these passages and turbines rotate due to this flow of water under head of water. Thus, electricity is created using the turbines.

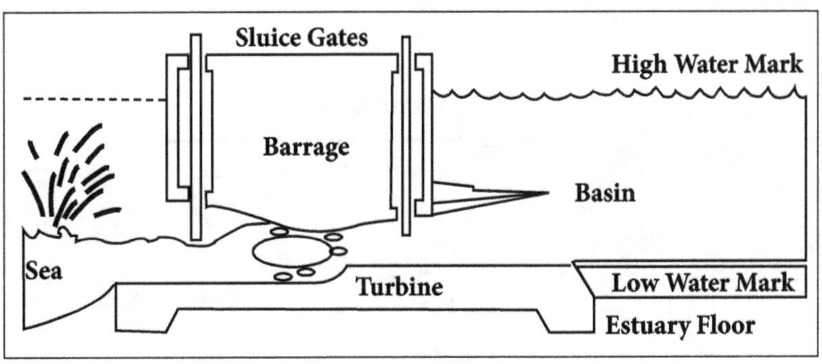

(a) General scheme of the tidal power station.

The components of a tidal power station are:

- Barrage: A barrage is a small wall built at the entrance of a gulf in order to trap the water behind it. It will either trap it by keeping it from going into the gulf when the sea water level is high or it will keep water from going into the sea when the sea water level is low.
- Turbines: These are the components responsible for converting the potential energy into kinetic energy. It is located in the passageways that the water flows through when the gates of the barrage are opened. There are many types of turbines used in tidal power stations such as:
    - Bulb turbines: As shown in figure (b). These are difficult to maintain as water flows around them and the generator is in water.
    - Rim turbines: As shown in figure (c). These are better maintained than the bulb turbines but are hard to regulate as generator is fixed in barrage.
    - Tabular turbines: As shown in figure (d). These turbines are fixed to long shafts and thus solve both problems that bulb and rim turbines have as they are easier to maintain and control.

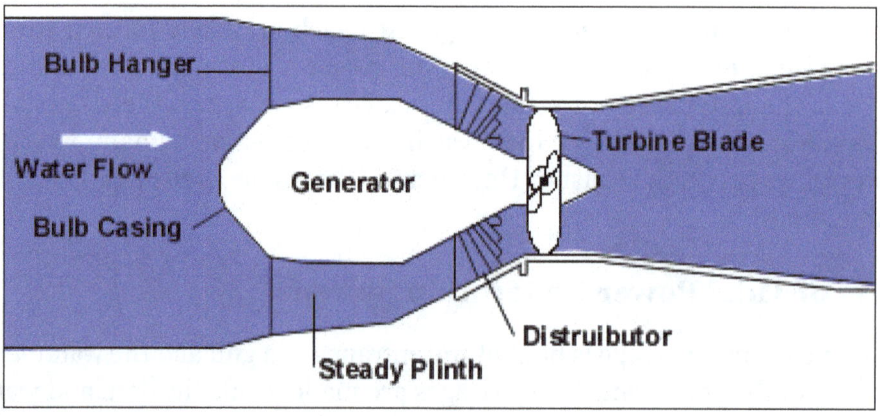

(b) A Bulb turbine.

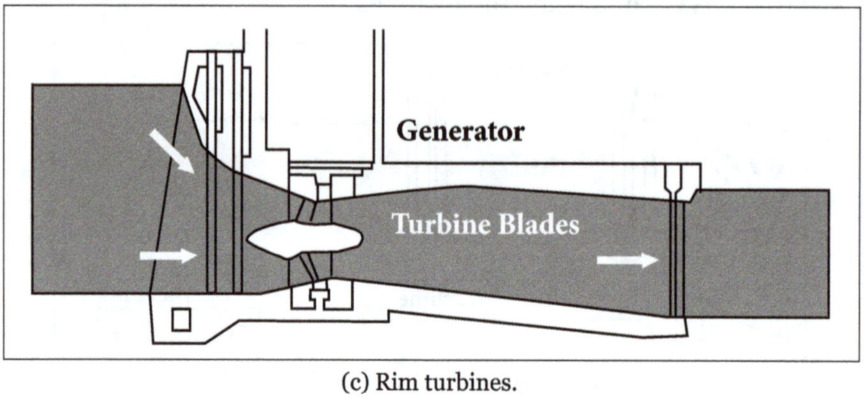

(c) Rim turbines.

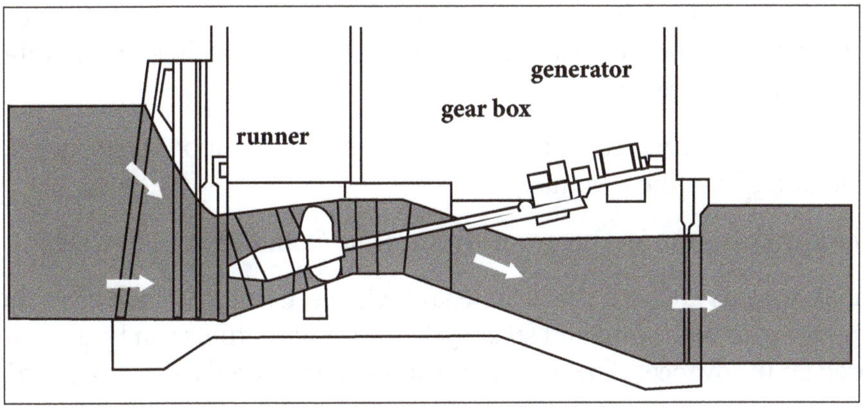

(d) Tabular turbines.

- Sluices: They are responsible for the flow of water through the barrage.
- Embankments: They are caissons made out of concrete to prevent the water from flowing at a certain parts of the dam and to help the maintenance work and the electrical wiring to be connected or used to move the equipment or cars over it. These embankments are shown in figure (e).

(e) Embankments.

Advantages of Tidal Power Generation

There are many advantages of generating power from the tide such as:

- Tidal power is a renewable and sustainable energy resource.
- It produces no liquid or solid pollution.
- It reduces dependence upon fossil fuels.
- Construction of large-scale offshore devices results in new areas of sheltered water, attractive for fish, sea birds, seals and seaweed.
- It has little visual impact.

- The tidal turbine offers significant environmental advantages over wind and solar systems, the majority of the assembly is hidden below the waterline and all cabling is along the seabed.
- It offers short time scale between investing in the modular construction and benefiting from the revenue.
- The tidal power exists on a worldwide scale from deep ocean waters.
- Tidal currents are both predictable and reliable, a feature which gives them an advantage over both wind and solar systems. Power outputs can be accurately calculated far in advance, allowing for easy integration with the existing electricity grids.
- Tidally driven coastal currents provide an energy density four times greater than air, meaning that a 15-m diameter turbine will generate as much energy as a 60mdiameter windmill.

Disadvantages and Constraints to Tidal Power Generation

It also has many disadvantages and limitations to generating tidal power. Some of them are:

- Tidal fences could present some difficulty to migrating fish.
- It does not generate electricity at a steady rate and thus not necessarily at times of peak demand, so unless a way can be found of storing energy efficiently and any storage devices currently available incur a considerable loss as they would not help in reducing the overall need for the fossil power stations, but only allow them to run at a lower rating for a certain amount of the time.
- At the present time both the tide and the wave energy are suffering from orientation problems, in the sense that neither method is strictly economical on a large scale in comparison with the conventional power sources.

Tidal Energy

The rise and fall of the sea level can power electric generating equipment. The gearing of equipment is tremendous to turn the very slow motion of the tide into enough displacement to generate energy.

The tidal barrages built across suitable estuaries are designed to obtain energy from the rise and fall of the tides using turbines located in the water passages in the barrages.

The potential energy due to the various level of water is across barrages is converted into kinetic energy in the form of fast moving water passing through the turbines. This in turn is converted into rotational kinetic energy by the blades of the turbine then driving a generator to produce electricity. The diagram demonstrates the power generation cycle of a tidal power.

The highest output can be achieved from the hydroelectric turbines by operating when the available head is the highest.

The available head is highest at extreme low tide and extreme high tide. These periods are roughly two hours in length, but there is relatively little change in the water level during half hour preceding and the next half hour after each of the extreme lows and highs. By including these 30 minute "shoulder" periods, a three hour generation period can be achieved twice per tidal cycle.

Thus, one can effectively generate at optimum levels for roughly half of each tidal cycle. Tidal cycles do not correspond to daily cycles of demand for electricity.

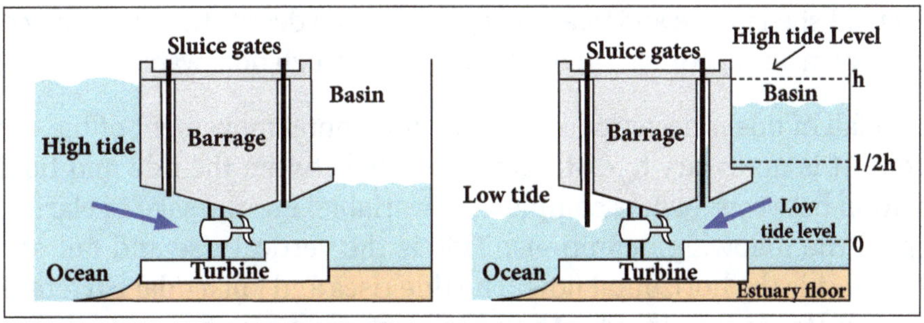

Power Generation from a Tidal Power.

Double Basin Arrangement

A double basin scheme can provide power continuously or on demand which is a great advantage. In the simplest double-basin scheme, there must be a dam between each basin and sea.

Working: In this type, the two basins are located apart and their waters are never exchanged.

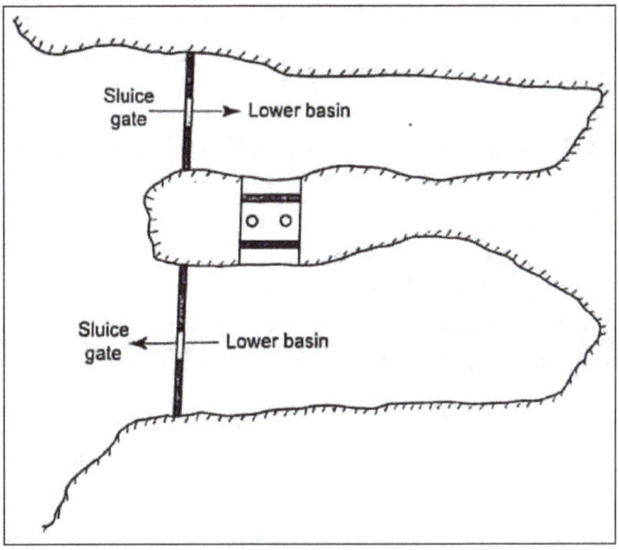

Tidal-power plant double basin.

The turbine is set up between the two basins. One basin in intermittently filled by the other is intermittently drained by the ebb tide. Water flow from high basin to low basin through turbines.

Movement of Tides and their Effect on Sea Level

Tides are the periodic motion of the waters of the sea due to changes in attractive forces of the Moon and Sun upon the rotating Earth. Tides can either help or hinder a mariner. A high tide may provide enough depth to clear a bar, while a low tide may prevent entering or leaving a harbor. The tidal current may help progress or hinder it, may set the ship toward dangers or away from them. By understanding tides and making intelligent use of predictions published in tide and tidal current tables and descriptions in sailing directions, navigator can plan an expeditious and safe passage through tidal waters.

The rise and fall of tide is accompanied by the horizontal movement of the water called tidal current. It is necessary to distinguish clearly between the tide and tidal current, for the relation between them is complex and variable. For the sake of clarity mariners have adopted the following definitions: Tide is the vertical rise and fall of the water and tidal current is the horizontal flow. The tide rises and falls, tidal current floods and ebbs. The navigator is concerned with the amount and time of the tide, as it affects access to shallow ports. The navigator is concerned with the time, speed and direction of tidal current, as it will affect his ship's position, speed and course. The tides are superimposed on nontidal rising and falling water levels, caused by weather, seismic events or other natural forces. Similarly, tidal currents are superimposed upon non-tidal currents such as normal river flows, floods and freshets.

6.6 Ocean Thermal Energy

Ocean Energy

Ocean can also be used for generating energy in the following ways.

Ocean Thermal Energy

There is often a large temperature difference between the surface level and deeper level of the tropical oceans. This temperature difference can be utilized to generate electricity. The energy available due to the difference in temperature of water is called thermal energy.

Condition

The temperature difference should be 20°C or more is required between surface water and ground water.

Process

The warm surface water of ocean is used to boil a low boiling liquid like ammonia. The high vapor pressure of the generator and generates electricity. The cold water from the deeper ocean is pumped to cool and condense the vapor into liquid.

Significance of OTE

- OTE is continuous, renewable and pollution free.
- The use of cold deep water as the chiller fluid in air-conditioning, has also been proposed.
- Electric power generated by OTE can be used to produce hydrogen.

Cycle Types

Cold seawater is an integral part of each of the three types of OTEC systems: closed-cycle, open-cycle and hybrid. To operate, the cold seawater must be brought to the surface. Primary approaches are active pumping and desalination. Desalinating seawater near the sea floor lowers its density, which causes it to rise to the surface.

The alternative to costly pipes to bring condensing cold water to the surface is to pump vaporized low boiling point fluid into the depths to be condensed, thus reducing pumping volumes and reducing technical and environmental problems and lowering costs.

Closed Cycle

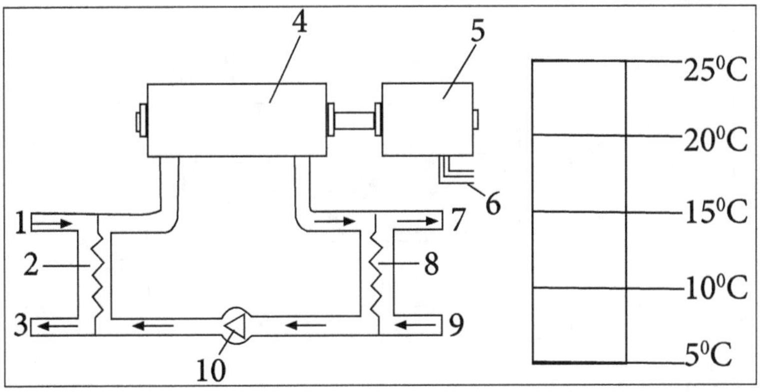

Diagram of a closed cycle OTEC plant.

- Surface water - 25°C.
- Evaporator.
- Waste water - 23°C.
- Turbine.

- Generator.
- Line to the grid.
- Waste water - 7°C.
- Condenser.
- Deep water - 5°C.
- Circulation pump.

Closed-cycle systems use fluid with a low boiling point, such as ammonia, to power a turbine to generate electricity. The warm surface seawater is pumped through a heat exchanger to vaporize the fluid. The expanding vapor turns the turbo-generator. The cold water is pumped through a second heat exchanger, condenses the vapor into a liquid, which is then recycled through the system.

Natural Energy Laboratory and several private sector partners developed the "mini OTEC" experiment, which achieved the first successful at sea production of net electrical power from closed-cycle OTEC. The mini OTEC vessel was moored 1.5 miles (2 km) off the Hawaiian coast and produced enough net electricity to illuminate the ship's light bulbs and run its computers and television.

Open Cycle

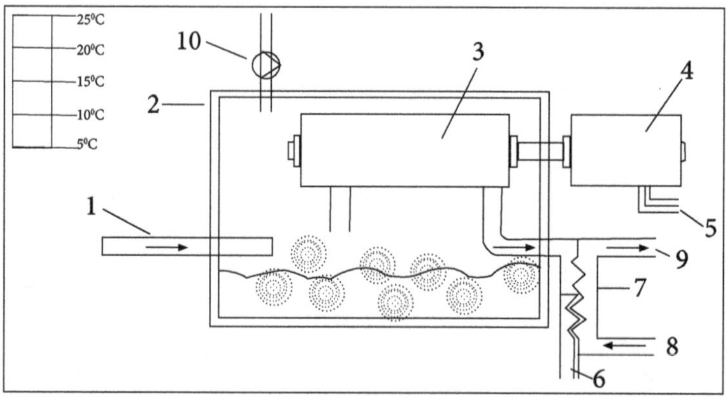

Diagram of an open cycle OTEC plant.

- Surface water - 25°C.
- Vacuum chamber, 3% to 1% of atmospheric pressure.
- Turbine.
- Generator.
- 5 Line to the grid.

- Desalinated water - 23°C.
- Condenser.
- Deep water - 5°C.
- Waste water - 7°C.
- Vacuum pump.

Open-cycle OTEC uses warm surface water directly to make the electricity. Placing the warm seawater in a low-pressure container causes it to boil. The expanding steam drives a low pressure turbine which is attached to an electrical generator. The steam, which has left its salt and other contaminants in the low-pressure container, is pure fresh water. It is condensed into a liquid by exposure to the cold temperatures from deep ocean water. This method produces desalinized fresh water, suitable for drinking water or irrigation.

Hybrid Cycle

It combines the features of the closed- and open-cycle systems. In a hybrid, warm seawater enters a vacuum chamber and is flash evaporated, similar to the open-cycle evaporation process.

The steam vaporizes ammonia working fluid of a closed-cycle loop on the other side of an ammonia vaporizer. The vaporized fluid then drives a turbine to produce electricity. The steam condenses within the heat exchanger and provides desalinated water.

Working Fluids

A popular choice of working fluid is ammonia, which has superior transport properties, easy availability and low cost. Ammonia is toxic and flammable. Fluorinated carbons such as CFCs and HCFCs are not toxic or flammable, but they contribute to the ozone layer depletion. Hydrocarbons are good candidates, but they are highly flammable: in addition, this would create competition for use of them directly as fuels. The power plant size is dependent upon the vapor pressure of working fluid. With increasing vapor pressure, the size of the turbine and heat exchangers decreases while the wall thickness of pipe and heat exchangers increase to endure high pressure especially on the evaporator side.

Schematic Diagram and Explanation

Ocean thermal energy conversion (OTEC) uses the difference between cooler deep and warmer shallow or surface ocean waters to run a heat engine and produce useful work, usually in the form of electricity.

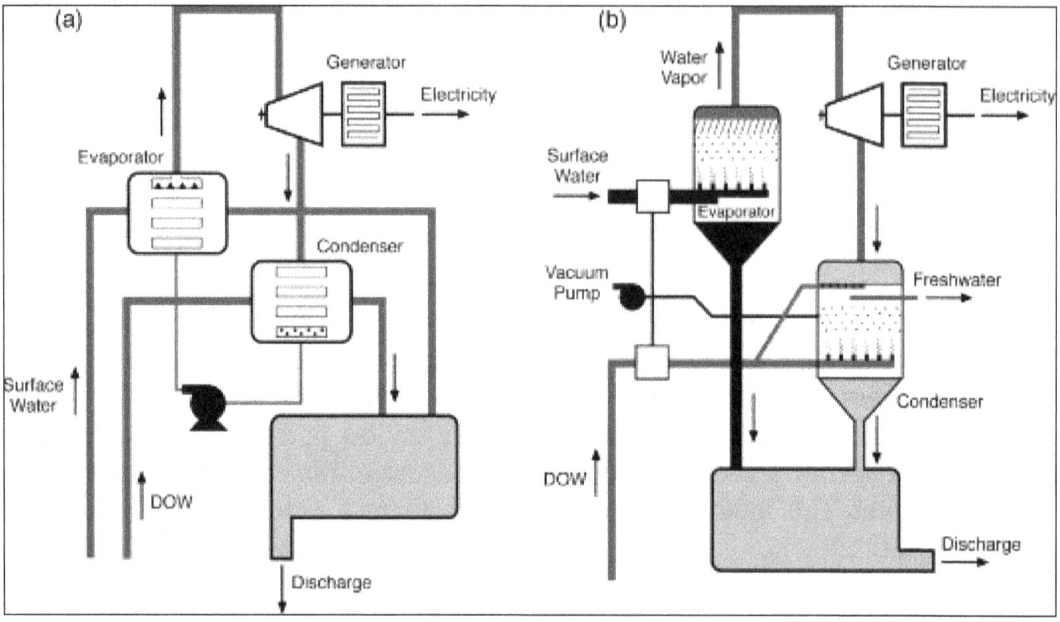

Ocean thermal energy conversion.

A heat engine gives greater efficiency and power when run with a large temperature difference. In the oceans, temperature difference between surface and deep water is greatest in the tropics, however still a modest 20°C to 25°C. It is therefore in the tropics that OTEC offers the greatest possibilities. The OTEC has the potential to offer global amounts of energy that are 10 to 100 times greater than other ocean energy options such as wave power. OTEC plants can operate continuously providing a base load supply for an electrical power generation system.

The main technical challenge of the OTEC is to generate significant amounts of power efficiently from small temperature differences. It is still considered an emerging technology. Early OTEC systems were of 1 to 3% thermal efficiency, well below the theoretical maximum for this temperature difference of between 6 and 7%. Current designs are expected to be closer to the maximum. The first operational system was built in Cuba in 1930 and generated 22 kW. Modern designs allow performance approaching the theoretical maximum Carnot efficiency and the largest built in 1999 by the USA generated 250 kW.

The most commonly used heat cycle for OTEC is the Rankine cycle using a low pressure turbine. Systems may be either closed-cycle or open-cycle.

The closed cycle engines use working fluids that are typically thought of as refrigerants such as ammonia or R-134a. The open-cycle engines use vapour from the seawater itself as the working fluid.

OTEC can also supply quantities of cold water as a byproduct. This can be used for air conditioning and refrigeration and the fertile deep ocean water can feed biological technologies. Another by-product is fresh water distilled from the sea.

Advantages

Ocean thermal energy is a renewable source of energy and it does not cause any pollution. Sea-wave energy and ocean thermal energy are the two forms of energy in which the solar energy manifests itself in oceans.

Limitation

The installation and maintenance of an OTEC power plant is very high and its efficient commercial exploitation is difficult.

6.7 Biomass and Biofuels

Biomass

Biomass is the organic matter, produced by plants or animals, used as source of energy; most of the biomass is burned directly for heating, cooling and industrial purposes.

Examples:

Wood, crop residues, seeds, cattle dung, sewage, agricultural wastes, etc. Biomass energies are of any one of the following types:

Advantages

- It is a renewable source of energy.
- It is a comparatively lesser pollution generating energy.
- Biomass energy facilitates in cleanliness in villages and cities.
- It gives manure for the agriculture and gardens.
- There is tremendous potential to generate biogas energy.
- The biomass energy is relatively cheaper and reliable.
- It will be generated from everyday human and animal wastes, vegetable and agriculture left-over etc.
- Recycling of waste decreases pollution and spread of diseases.
- Heat energy that one gets from the biogas is 3.5 times the heat from burning wood.
- Because of more heat produced the time needed for cooking is lesser.
- Pressure on the surrounding forest and scrubs will be reduced when biogas is used as cooking fuel.

- It is a more cost effective means of acquiring energy as compared to oil supplies. As oil supplies are getting depleted day by day, it is becoming a costly commodity.
- Growing biomass crops use up carbon dioxide and produces oxygen.

Disadvantages

- Cost of construction of biogas plant is high, so only rich people will use it.
- Continuous supply of biomass is needed to generate biomass energy.
- Some people don't like to cook food on biogas produced from sewage waste.
- Biogas plant needs space and produces dirty smell.
- Due to improper construction many biogas plants are working inefficiently.
- It is difficult to store biogas in cylinders.
- Transportation of biogas through pipe over long distances is difficult.
- Many easily grown grains like corn, wheat are being used to make ethanol. This will have bad consequences if too much of food crop is diverted for use as fuel.
- Crops which are used to produce biomass energy are seasonal and are not available over whole year.

Bio Gas

Bio gas is a mixture of gases such as methane, carbon dioxide, hydrogen sulfide, etc. It contains about 65% of methane gas as a major constituent.

Bio gas is obtained by the anaerobic fermentation of animal dung or plant wastes in the presence of water.

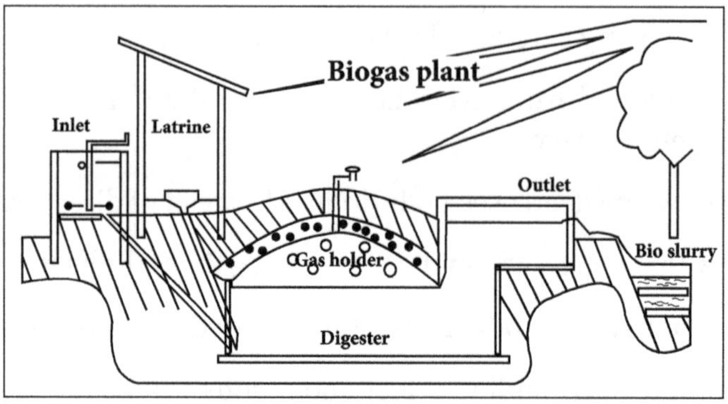

Biogas plant.

Bio Fuels

They are produced from living organisms or from metabolic by-products. In order to be considered a biofuel the fuel must contain over 80% renewable materials. It is originally derived from the photosynthesis process and can therefore often be referred to as a solar energy source. There are many pros and cons to using biofuels as an energy source.

The term biofuels refers to a wide range of fuels which are in some way derived from biomass organic matter. In today's society they are gaining increased public and scientific attention due to rising gas and oil prices. The most widely recognized biofuel is likely corn or sugar based ethanol. This is created through the fermentation of plant products. Bio-diesel is another form of biofuel that is made from vegetable oils, animal fats or recycled greases. It is produced from the oils or fats through a process called trans-esterification. The Inter-nation Energy Agency predicts that biofuels have the potential to meet more than a quarter of world demand for transportation. Bio fuels are the fuels, obtained by the fermentation of bio mass.

Examples: Ethanol, Methanol.

1. Ethanol

Ethanol can be produced from sugarcane. Its calorific value is less when compared to petrol and produces much less heat than petrol.

2. Methanol

It can be easily obtained from ethanol or sugar containing plants. Its calorific value is also too low when compared to gasoline and diesel.

3. Gasohol

Gasohol is a mixture of ethanol and gasoline. In India trial is being carried out to use Gasohol in cars and buses.

Hydrogen Fuel

Hydrogen can be produced by thermal dissociation or photolysis or electrolysis of water. It possess high calorific value. It is non-polluting.

$$2H_2 + O_2 \rightarrow 2H_2O + 150\ \text{kV}.$$

Disadvantages of Hydrogen Fuel

- Hydrogen is highly inflammable and explosive in nature.
- Safe handling is required.

- It is difficult to store and transport.

Advantages of Biofuels

Biofuels are a renewable energy source in that they are created from the plants that can be regrown each year. Biofuels also do not require many changes in cars and other places of use to be utilized. Some consider the use of biofuels as carbon neutral since the carbon produced, when burning them is offset by carbon consumed by the plants they came from. In the United States, biofuels can help reduce the dependence on foreign oils, which fluctuate in price rapidly. Biofuels may help to buffer against the change.

Disadvantages of Biofuels

While some consider their use "carbon neutral," the machinery required to farm the plants for biofuels does create carbon emissions, this machinery is also typically not powered by biofuels. Research suggests despite this fact that biofuels help to reduce carbon emissions by 50-60%.

One of the main detractors to the use of biofuels is that setting aside land for biofuel crops means less land for food production. Some foreign countries have said that it is unethical to use crops for biofuel when global hunger is an ever present problem.

Examples of Various Biofuels used in Engineering Applications

Bio-diesel

It is any kind of biofuel made from animal fats and vegetable oils, either new or recycled. It is the most common biofuel in Europe and is produced mainly from soy, sunflower, palm oil and jatropha but also from other vegetables like rapeseed, mahua, mustard, flax and even algae.

Bio-ethanol

It is most common biofuel worldwide. It is a kind of bio-alcohol and is produced by the fermentation of sugars contained in wheat, corn, sugar beets, sugar cane, etc. Although, sugars are not the only way to produce bio-ethanol cellulose is another raw material used in the production of second generation biofuels, though it is more difficult to process. HRS Heat Exchanger has developed its own process for enhancing the production of cellulosic ethanol.

Biogas

It receives the methane produced through anaerobic digestion of organic material. The solid by-product, called digestate can be used as biofuel or fertilizer. There are several subprocess and steps that can be enhanced with our technologies such as exhaust gas cooling, sludge heating and even digestate evaporation at the end of the biogas production process.

6.8 Fuel Cells: Cell Representation and H_2-O_2 Fuel Cell: Design and Working, Advantages and Limitations

Full Cell

A full cell converts the chemical energy of the fuels directly to electricity. The essential process in a fuel cell is,

Fuel + Oxygen $\rightarrow$ Oxidation products + Electricity.

Fuel cells work like batteries, but they do not run down or need recharging. They produce electricity and heat as long as fuel is supplied. A fuel cell consists of two electrodes, a negative electrode (anode) and a positive electrode (cathode) - sandwiched around an electrolyte. A fuel, such as hydrogen, is fed to anode and air is fed to the cathode. In a hydrogen fuel cell, a catalyst at the anode separates the hydrogen molecules into protons and electrons, which take different paths to the cathode. The electrons go through an external circuit, creating a flow of electricity. The protons migrate through the electrolyte to the cathode, where they unite with the oxygen and electrons to produce water and heat.

Cell Representation

A fuel cell can be represented as Fuel Electrode | Electrolyte | Electrode | Oxidant.

The common fuels are H_2, N_2H_4 and Hydrocarbons. The major oxidants are Oxidants O_2, H_2O_2, HNO_3, etc.

A fuel cell is a galvanic cell in which the chemical energy associated with the oxidation of reducing agents (fuels) is directly converted into electrical energy. The conventional method of utilizing the chemical energy of fuel to produce electrical energy and the direct conversion process used in the fuel cell can be represented as shown below.

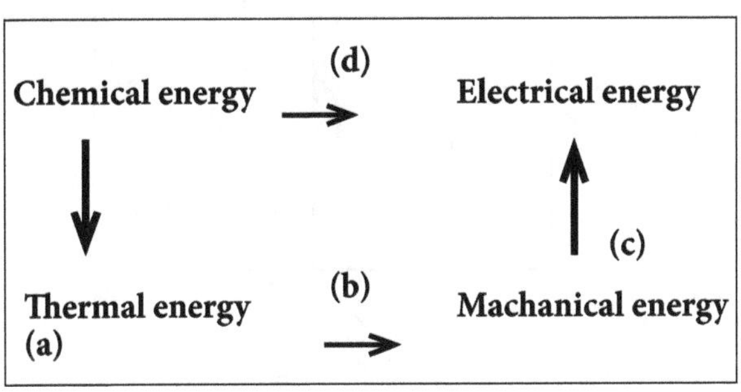

Any losses in energy in steps (a) and (c) can be minimized. Although the efficiency of the process (b) is limited by the second law of thermodynamics. Hence the process (d) might be expected to have a higher efficiency. If ΔH is the enthalpy change of the reaction, the amount of useful work that can be obtained out of this is ΔG. The rest of it. i.e., ΔH - ΔG = T ΔS is unavailable for work.

Efficiency = E = Work obtained from the cell reaction / Heat change accompanying the reaction

$$= \Delta G / \Delta H$$

$$E = \Delta G / H = 1 - T\left(\Delta S / \Delta H\right)$$

Thus, E depends on both ΔS and ΔH.

A typical example of pollution free cell is $H_2 - O_2$ fuel cell in which the fuel is hydrogen and oxidizer is oxygen.

Advantages of Fuel Cells

- Fuel cells are efficient and takes less time for operation.
- No harmful chemicals are produced in fuel cells.

Uses

- It is used as auxiliary energy source in space vehicles, submarines etc.
- It is used in producing drinking water for astronauts in the space.

$H_2 - O_2$ fuel cell

Design and Working

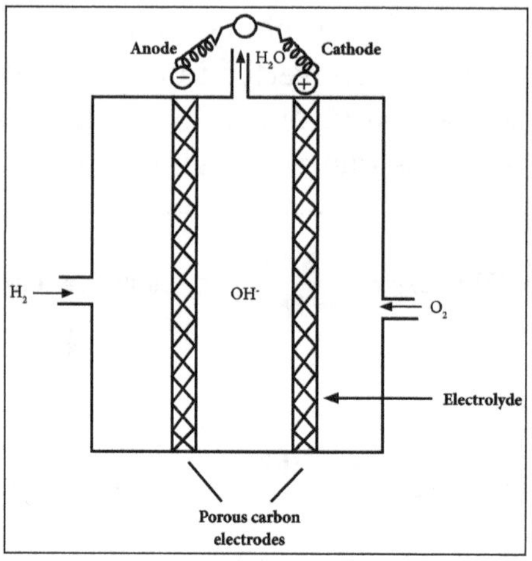

H_2- O_2 cell.

Hydrogen - Oxygen fuel cell consists of two porous electrodes made up of compressed carbon coated with small amount of catalysts (P_t, P_d, A_g) and KOH or NaOH solution as the electrolyte.

During working, hydrogen (the fuel) is bubbled though the anode compartment where it is oxidized. The oxygen (oxidizer) is bubbled through the cathode compartment where it is reduced. The following cell reactions occur.

Anodic reaction: $2H_2 + 4OH^- \rightarrow 4H_2O + 4e^-$.

Cathodic reaction: $O_2 + 2H_2O + 4e^- \rightarrow 4OH^-$.

Overall cell reaction: $2H_2(g) + O_2(g) \rightarrow 2H_2O(l)$.

From the above cell reactions, the hydrogen molecules are oxidized to water. When a large number of fuel cells are connected in series it is called fuel battery.

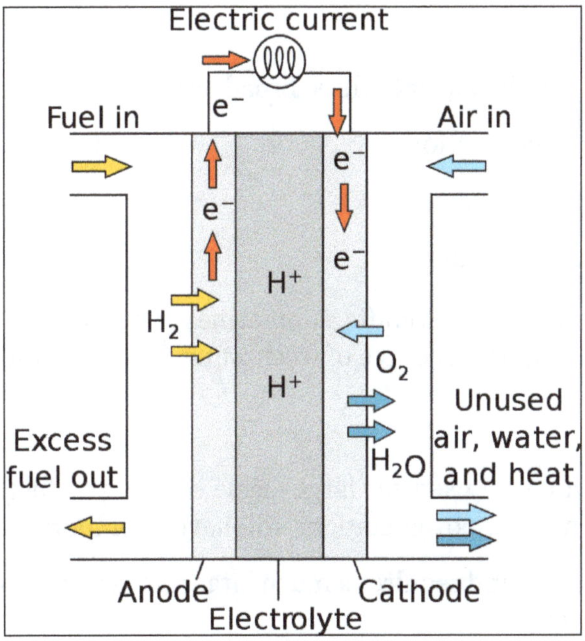

H_2- O_2 Fuel cell.

A fuel cell is like a battery. It has two electrodes where the reactions take place and an electrolyte which carries the charged particles from one electrode to the other. In order for a fuel cell to work, it needs the hydrogen (H_2) and oxygen (O_2).

The hydrogen enters the fuel cell at the anode. A chemical reaction strips hydrogen molecules of their electrons and the atoms become ionized to form H^+.

The electrons travel through wires to provide a current to do work. The oxygen enters at the cathode usually from the air. The oxygen picks up the electrons that have completed their circuit. The oxygen then combines with the ionized hydrogen atoms (H^+)

and water (H_2O) is formed as the waste product which exits the fuel cell. The electrolyte plays an essential role as well. It only allows appropriate ions to pass between the anode and cathode. If other ions were allowed to flow between anode and cathode, the chemical reactions within the cell would be disrupted.

The reaction in a single fuel cell typically produces only about 0.7 volts can be used in cars, generators or other products that require power.

The reactions involved in a fuel cell are:

Anode side (an oxidation reaction): $2H_2 \rightarrow 4H^+ + 4e^-$.

Cathode side (a reduction reaction): $O_2 + 4H^+ + 4e^- \rightarrow 2H_2O$.

Net reaction: $2H_2 + O_2 \rightarrow 2H_2O$.

Applications

The use of fuel cells mainly falls into three broad areas:

- Portable power generation.
- Stationary power generation.
- Power for transportation.

$H_2 - O_2$ fuel cells is used in space crafts, submarines to get electricity. In H_2-O_2 fuel cell, the product water is a valuable source of fresh water for astronauts.

Limitations

- Fuel cells cannot be used for large–scale energy production, so conventional fossil fuel or nuclear power stations still have an important future.
- Hydrogen is a gas and requires a much larger storage volume compared to fossil fuels like petrol.
- Safe storage is an issue, especially as it would be stored under high pressure to decrease the storage space required.
- This immediately makes leaks and accidents more likely to happen because the hydrogen is a highly flammable explosive gas – too easily ignited.
- There is no efficient means of mass producing hydrogen.
- Efficient large scale technology is not yet developed to produce hydrogen on a large scale example from electrolysis using solar power electricity – photovoltaic power system, wind turbines or hydroelectric power.

- However water is cheap and plentiful, it requires expensive electrical energy to electrolyze water to split it into hydrogen and oxygen.
- The electrolysis of acidified water is expensive because electricity is expensive and much is still made by the burning fossil fuels. Not only that, most electricity in the world is generated from burning fossil fuels.

6.9 Types of Fuel Cells: Alkaline Fuel Cell, Methanol-Oxygen, Phosphoric Acid Fuel Cells and Molten Carbonate Fuel Cells

Alkaline Fuel Cell

Alkaline fuel cells (AFCs) were one of the first fuel cell technologies developed and they were the first type broadly used in the U.S. space program to produce electrical energy and water on-board spacecraft. These fuel cells use a solution of potassium hydroxide in water as the electrolyte and can use a variety of non-precious metals as a catalyst at anode and cathode. In recent years, novel AFCs that use a polymer membrane as the electrolyte have been developed. These fuel cells are similar to conventional PEM fuel cells, except that they use an alkaline membrane instead of an acid membrane. The high performance of AFCs is due to the rate at which electro-chemical reactions take place in cell. They have also demonstrated efficiencies above 60% in space applications.

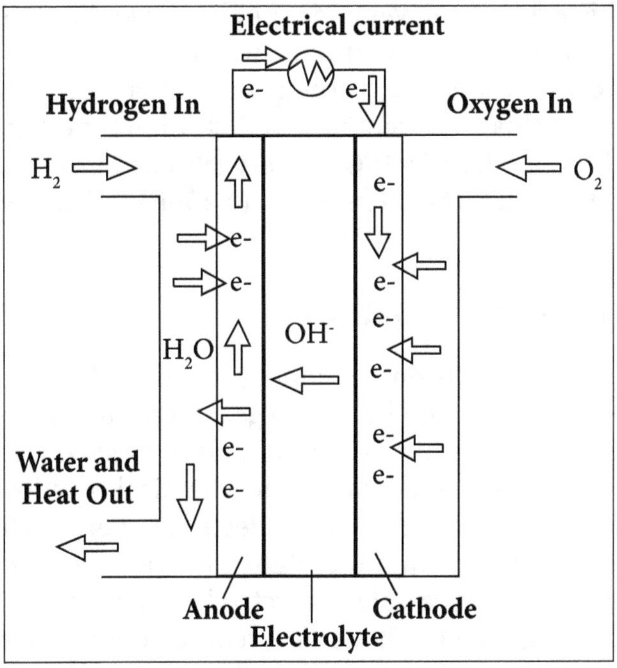

Alkaline fuel cell.

A key challenge for this fuel cell type is that it is susceptible to poisoning by carbon dioxide (CO_2). In fact; even the small amount of CO_2 in air can dramatically affect cell performance and durability due to carbonate formation. Alkaline cells with liquid electrolytes can be run in a recirculating mode, which allows for electrolyte regeneration to help reduce the effects of carbonate formation in electrolyte, but the recirculating mode introduces issues with shunt currents. The liquid electrolyte systems also suffer from additional concerns including wettability, increased corrosion and difficulties handling differential pressures. The alkaline membrane fuel cells (AMFCs) address these concerns and have lower susceptibility to CO_2 poisoning than liquid-electrolyte AFCs. However, CO still affects performance and durability of the AMFCs still lag that of PEMFCs. The AMFCs are being considered for applications in the W to kW scale. Challenges for AMFCs include tolerance to carbon dioxide, membrane conductivity and durability, higher temperature operation, water management, power density and anode electro catalysis.

Methanol Oxygen Fuel Cells

In a fuel cell electrical energy is obtained from oxygen and a fuel that can be oxidized. The essential process in a fuel cell is,

$$\text{Fuel} + \text{Oxygen} \rightarrow \text{Oxidation product} + \text{Electricity}$$

Methanol – oxygen fuel cell are powered by pure methanol. In the early 1960s, the mechanism of this reaction was the subject of many studies in different countries. Still, the specific power attained was quite small. Only in the mid-1990s, after the great successes achieved in the development of hydrogen-oxygen fuel cells with the proton-conducting ion exchange membrane, was a breakthrough reached in the development of fuel cells with direct (without preliminary conversion) oxidation of methanol. Fuel cells of this type were now called "direct-methanol fuel cells"(DMFC). The design of the modern DMFCs is very similar to the design of proton exchange-membrane fuel cells. The membranes (Nafion) and catalysts (Platinum) are used.

Direct Methanol Fuel Cell (DMFC)

It is a fuel cell that runs directly on methanol (or various liquid fuels) without having to first convert those fuels into hydrogen gas. It consists of two electrodes separated by a proton exchange membrane (PEM) and connected via an external circuit that allows conversion of free energy from the chemical reaction of methanol with air or oxygen to be directly converted into electrical energy. The direct methanol fuel cell is a proton exchange membrane (PEM) fuel cell that is fed with an aqueous solution of methanol. The two catalytic electrodes where the methanol oxidation (anode) and oxygen reduction (cathode) are separated by a membrane which conducts protons from anode to cathode, while other compounds diffusion is blocked.

The combination of electrodes and membranes is called membrane electrode assembly (MEA). MEA was prepared with Pt/Ru black at anode and Pt black cathode on either side of a Nafion membrane. Each electrode is made of a gas diffusion layer and a catalytic layer.

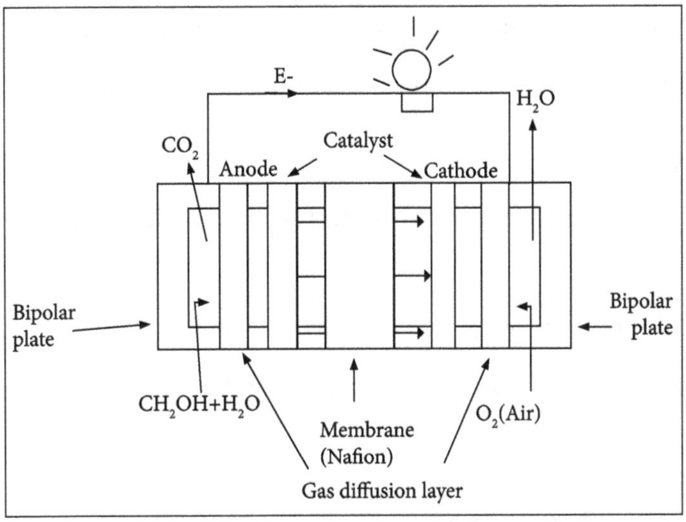

Direct Methanol – Oxygen Fuel Cell (DMFC).

Membrane Nafion was created by the addition of sulfonic acid groups into the bulk polymer matrix of Teflon. These sites have strong ionic properties and act as proton exchange sites. The aqueous methanol is fed at the anode side. It diffuses through the diffusion layer to catalytic layer where it is electrochemically oxidized into mainly carbon dioxide, protons and electrons. Protons formed during this reaction diffuse through the Nafion membrane to the cathode catalytic layer. They participate in oxygen reduction to form water at the cathode side. Oxygen may be pure but can also come from air. Electrons are collected by the graphite bipolar plates which are the two poles of the cell.

Reactions

Anode reaction: $CH_3OH + 7H_2O \rightarrow CO_2 + 6H_3O^+ + 6e^-, E^o = 0.02V$

Cathode reaction: $3/2O_2 + 6H_3O^+ + 6e^- \rightarrow 9H_2O, E^o = 1.23V$

Overall reaction: $CH_3OH + 3/2O_2 \rightarrow CO_2 + 2H_2O, E^o = 1.21V$

At anode, methanol is oxidized into carbon dioxide and six protons (as hydronium ions) plus six electrons. The six protons formed react at the cathode with oxygen to form water. The overall reaction looks like a combustion reaction and is thus sometimes referred tossss as cold combustion. Actually the cell is a mean to control this reaction and use it to produce current directly. The standard cell voltage for a DMFC at 25°C is 1.21V. However, this potential is never obtained in reality. The open circuit potential is usually about 0.6 to 0.8V.

Advantages

- The methanol can be refueled quickly and simply.
- The methanol is arguably the most efficient carrier of hydrogen.
- The simplicity of methanol storage drastically reduces the weight of the system.
- Ideal for low power-density, but high energy-density applications.

Disadvantages

- The fuel can cross electrolyte without reacting, reducing efficiency.
- Low power-density.
- The methanol can corrode various parts of the fuel cell.
- The methanol is toxic and flammable, much like gasoline.

Applications

A potential area of application for DMFCs is low-power (up to 20 W) power sources for electronic equipment, such as notebooks, video cameras, DVD-players, cell phones, medical devices, etc. At this time, the application of DMFCs as power sources for the electric vehicles is very remote.

Substantial progresses are also made in the design and prototyping of the DMFC stacks for portable and auxiliary power applications. Three generations of highly efficient and lightweight 11-W stacks were developed for the operation at a design voltage of 0.55/ cell. Four units from the last generation were supplied to Aerospace and Technologies Corporation and prototypes for the military are also available.

Phosphoric Acid Fuel Cells

It uses liquid phosphoric acid as an electrolyte the acid is contained in a Teflon-bonded silicon carbide matrix and porous carbon electrodes containing a platinum catalyst. The electro-chemical reactions that take place in the cell are shown in the diagram.

The PAFC is considered the "first generation" of the modern fuel cells. It is one of the most mature cell types and the first to be used commercially. This type of fuel cell is usually used for stationary power generation, but some PAFCs have been used to power large vehicles such as city buses.

The PAFCs are more tolerant of impurities in fossil fuels that have been reformed into hydrogen than PEM cells, which are easily "poisoned" by carbon monoxide because carbon monoxide binds to the platinum catalyst at the anode, decreasing fuel cell's efficiency. They are more than 85% efficient when used for the co-generation of electricity and heat but they are less efficient at generating electricity alone (37%–42%). PAFC efficiency is only slightly more than that of combustion-based power plants, which typically operate at around 33% efficiency. PAFCs are also less powerful than other fuel cells, given same weight and volume. As a result, these fuel cells are typically large and heavy. They are also expensive and require much higher loadings of expensive platinum catalyst than other types of fuel cells do, which raises the cost.

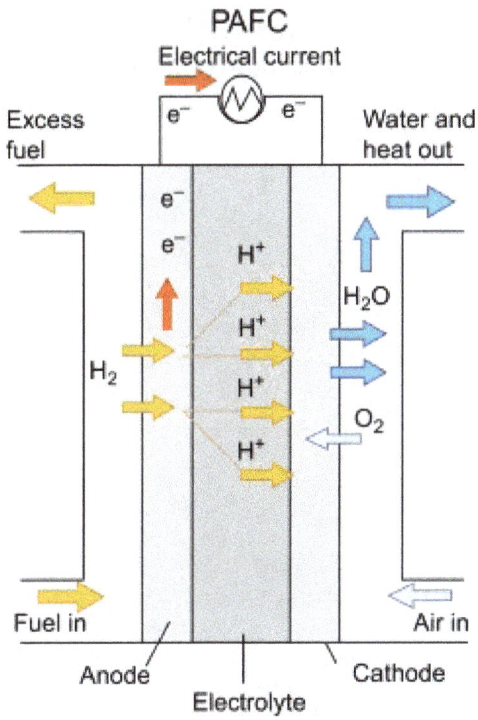

Phosphoric Acid Fuel Cells.

Molten Carbonate Fuel Cells

They are currently being developed for natural gas and coal-based power plants for electrical utility, industrial and military applications. They are high-temperature fuel cells that use an electrolyte composed of a molten carbonate salt mixture suspended in a porous, chemically inert ceramic lithium aluminum oxide matrix. Because they operate at high temperatures of 650°C (roughly 1,200°F), non-precious metals can be used as catalysts at the anode and cathode, reducing costs.

Improved efficiency is another reason the MCFCs offer significant cost reductions over phosphoric acid fuel cells. Molten carbonate fuel cells, when coupled with a turbine, can reach efficiencies approaching 65%, considerably higher than the 37%–42% efficiencies of a phosphoric acid fuel cell plant. When the waste heat is captured and used, overall fuel efficiencies can be over 85%.

Unlike alkaline, the phosphoric acid and PEM fuel cells, MCFCs do not require an external reformer to convert fuels such as natural gas and biogas to hydrogen. At the high temperatures at which MCFCs operate, methane and other light hydrocarbons in these fuels are converted to hydrogen within the fuel cell itself by a process called internal reforming, which also reduces cost.

The primary disadvantage of current MCFC technology is durability. The high temperatures at which these cells operate and corrosive electrolyte used accelerate component breakdown and corrosion, decreasing cell life. The scientists are currently exploring corrosion-resistant materials for components as well as fuel cell designs that double cell life from current 40,000 hours (~5 years) without decreasing performance.

Permissions

All chapters in this book are published with permission under the Creative Commons Attribution Share Alike License or equivalent. Every chapter published in this book has been scrutinized by our experts. Their significance has been extensively debated. The topics covered herein carry significant information for a comprehensive understanding. They may even be implemented as practical applications or may be referred to as a beginning point for further studies.

We would like to thank the editorial team for lending their expertise to make the book truly unique. They have played a crucial role in the development of this book. Without their invaluable contributions this book wouldn't have been possible. They have made vital efforts to compile up to date information on the varied aspects of this subject to make this book a valuable addition to the collection of many professionals and students.

This book was conceptualized with the vision of imparting up-to-date and integrated information in this field. To ensure the same, a matchless editorial board was set up. Every individual on the board went through rigorous rounds of assessment to prove their worth. After which they invested a large part of their time researching and compiling the most relevant data for our readers.

The editorial board has been involved in producing this book since its inception. They have spent rigorous hours researching and exploring the diverse topics which have resulted in the successful publishing of this book. They have passed on their knowledge of decades through this book. To expedite this challenging task, the publisher supported the team at every step. A small team of assistant editors was also appointed to further simplify the editing procedure and attain best results for the readers.

Apart from the editorial board, the designing team has also invested a significant amount of their time in understanding the subject and creating the most relevant covers. They scrutinized every image to scout for the most suitable representation of the subject and create an appropriate cover for the book.

The publishing team has been an ardent support to the editorial, designing and production team. Their endless efforts to recruit the best for this project, has resulted in the accomplishment of this book. They are a veteran in the field of academics and their pool of knowledge is as vast as their experience in printing. Their expertise and guidance has proved useful at every step. Their uncompromising quality standards have made this book an exceptional effort. Their encouragement from time to time has been an inspiration for everyone.

The publisher and the editorial board hope that this book will prove to be a valuable piece of knowledge for students, practitioners and scholars across the globe.

Index

www.ingramcontent.com/pod-product-compliance
Lightning Source LLC
LaVergne TN
LVHW062233200926
848484LV00006B/1533